The Plasma Universe

Plasma physics is the fascinating science behind lightning bolts, fluorescent lights, solar flares, ultra-bright TV screens, fusion reactors, cosmic jets, and black hole radiation, to name but a few examples. Research into plasma phenomena could lead to a source of unlimited, non-polluting energy. Yet plasmas obey their own, often very surprising, rules, and repeatedly defy our best efforts to anticipate and control them.

This richly illustrated full color book reveals for the first time the exciting world of plasma physics to a non-technical audience. It describes the phenomena, and follows the worldwide research effort to comprehend them, taking the reader on a journey from neighborhood neon lights to the remotest galaxies and beyond.

The lively writing is interspersed with fascinating photographs and explanatory diagrams, giving readers a deeper understanding of the world around them.

CURT SUPLEE is a renowned science writer. He was a writer and editor at *The Washington Post*, where he was twice nominated for a Pulitzer Prize. He has won the Science Journalism Award from the American Association for the Advancement of Science, the American Chemical Society's Grady–Stack Award for Interpreting Chemistry to the Public, and two national prizes from the American Astronomical Society.

THE PLASMA UNIVERSE

CURT SUPLEE

Edited by Amitava Bhattacharjee
Designed by Kristi Donahue

CAMBRIDGE UNIVERSITY PRESS
Cambridge, New York, Melbourne, Madrid, Cape Town, Singapore, São Paulo, Delhi

Cambridge University Press
The Edinburgh Building, Cambridge CB2 8RU, UK

Published in the United States of America by Cambridge University Press, New York

www.cambridge.org
Information on this title: www.cambridge.org/9780521519274

© The Division of Plasma Physics of the American Physical Society 2009

This publication is in copyright. Subject to statutory exception
and to the provisions of relevant collective licensing agreements,
no reproduction of any part may take place without the written
permission of Cambridge University Press.

First published 2009

Printed in the United Kingdom at the University Press, Cambridge

A catalogue record for this publication is available from the British Library

ISBN   978-0-521-51927-4 paperback

Cambridge University Press has no responsibility for the persistence or
accuracy of URLs for external or third-party Internet websites referred to in
this publication, and does not guarantee that any content on such websites is,
or will remain, accurate or appropriate.

About our cover image

The cover image was provided courtesy of Walter Gekelman, Stephen Vincena, and Andrew Collette of the Department of Physics and Astronomy, University of California, Los Angeles.

The image is an experimental result of measuring the magnetic field in three dimensions (40 000 spatial locations, every 10 nanoseconds), which happens when two dense plasmas (produced when lasers strike targets) collide. The collision takes place in a magnetized background plasma that supports Alfvén waves. The waves set up a 3-D current system, which, as it turns out, are currents of Alfvén waves.

Seen at the top is a magnetic "X" point which is undergoing magnetic field line reconnection. The reconnection results in an electric field which is rendered as glowing sparklers. The length of the sparkler is proportional to the electric field strength at that point.

Contents

List of Figures	page ix
Preface	xi
Chapter 1: The Fourth State of Matter	1
The Same But Different	5
What's In a Name?	9
Irving Langmuir	10
Forces and Feedbacks	12
Chapter 2: The Music and Dance of Plasmas	15
Langmuir Waves (plasma oscillations)	16
Alfvén Waves Down the Line	20
A Rough Ride from the Sun	24
Hannes Alfvén	25
Chapter 3: The Sun–Earth Connection	27
Lighting Up the Solar System	28
Spots, Loops, and Lariats of Fire	29
New Views of the Sun	31
Blast from the Mass	32
A Mighty Wind	34
Eugene N. Parker	38
Chapter 4: Bringing the Sun to Earth: The Story of Controlled Thermonuclear Fusion	39
Magnetic Bottles	41
Fusion by Light	46
Just a Pinch	49
Marshall Rosenbluth	50

Chapter 5: The Cosmic Plasma Theater:
Galaxies, Stars, and Accretion Discs 51
Shaping Up 52
Going to Extremes 54
Discs and Holes 56
Jet Propulsion 59
In the Firing Line 60

Chapter 6: Putting Plasmas to Work 63
A Plasma to Read By 64
Walls of Light 66
Withdrawal and Deposits 67
Plasmas and Human Health 70
When Push Comes to Shove 71

Index 73

Figures

Figure 1. Industrial plasma — 2
Image courtesy of Randy Montoya, Sandia National Laboratory

Figure 2. Auroral lights — 3
Image courtesy of www.dreamstime.com

Figure 3. Examples of plasmas — 4
Magnetic confinement fusion image courtesy of Neil Calder, ITER; nebula, solar wind, and solar corona images courtesy of NASA; aurora, flames, fluorescent light, neon sign, and lightning images courtesy of www.dreamstime.com; inertial confinement image courtesy of Lawrence Livermore National Laboratory, the National Ignition Facility, and the Department of Energy

Figure 4. DIII-D magnetic-confinement fusion device — 6
Image courtesy of General Atomics

Figure 5. Hohlraum — 8
Image courtesy of Lawrence Livermore National Laboratory, the National Ignition Facility, and the Department of Energy

Figure 6. Evolution of 2-D elliptical electron vortex — 9
Image courtesy of Travis B. Mitchell; research conducted at the Department of Physics and Astronomy, University of Delaware, under the auspices of the Department of Energy and the National Science Foundation

Figure 7. Irving Langmuir — 10
Photo courtesy of Chemical Heritage Foundation Collection

Figure 8. The vacuum tube — 11
*Left illustration drawn after Albert W. Hull and Irving Langmuir, **Proceedings of the National Academy of Sciences, USA** (March 15, 1929). Right illustration modified from triode tube schematic, licensed to Creative Commons Attribution-Share Alike 2.5 Netherlands*

Figure 9. Plasma discharge as lightning — 13
Image courtesy of www.dreamstime.com

Figure 10. The "Z-pinch" inertial confinement device — 14
Image courtesy of Randy Montoya, Sandia National Laboratory

Figure 11. Plasma lamp — 15
Image courtesy of Luc Viatour, licensed under the GNU Free Documentation License using material from Wikipedia

Figure 12. Langmuir wave oscillation — 16
*Illustration adapted from F. Chen, **Introduction to Plasma Physics and Controlled Fusion** (New York: Plenum Press, 1984), p. 83*

Figure 13. Langmuir waves in the wake of a laser pulse — 17
Image courtesy of Nicholas Matlis and Michael Downer, University of Texas at Austin; Anatoly Maksimchuk and Victor Yanovsky, University of Michigan

Figure 14. Electron acceleration by wake — 19
*Image reprinted by permission from Macmillan Publishers Ltd: I. Blumenfeld et al., **Nature** 445 (2007), p. 742; copyright 2007*

Figure 15. Alfvén wave fields — 20
*Illustration adapted from F. Chen, **Introduction to Plasma Physics and Controlled Fusion** (New York: Plenum Press, 1984), p. 139*

Figure 16. Alfvén wave visualization — 22
Image courtesy of Walter Gekelman, Department of Physics and Astronomy, University of California, Los Angeles

Figure 17. Alfvén wave in Wendelstein stellarator — 23
Image courtesy of A. Weller

Figure 18. Alfvén waves in the chromosphere — 23
Image courtesy of T. J. Okamoto

Figure 19. Alfvén waves in the solar wind — 24
*Image courtesy of J. W. Belcher and L. Davis. This image was published by AGU: **Journal of Geophysical Research** 376 (1971), no. 16; copyright 1971*

Figure 20. Hannes Alfvén — 25
Photo from the Glasheen Collection, courtesy of Mandeville Special Collections Library, University of California, San Diego

Figure 21. Comet Hale-Bopp — 26
Image courtesy of Fred Espenak, www.MrEclipse.com

Figure 22. Cutaway of the Sun — 28
Image courtesy of SOHO (ESA & NASA) Consortium

Figure 23. "Magnetogram" from the satellite Hinode — 29
Image copyright NAOJ / JAXA / NASA / SFC / ESA

Figure 24. New views of the Sun — 31
Image copyright NAOJ / JAXA / NASA / SFC / ESA

Figure 25. Coronal loops — 32
Image courtesy NASA

Figure 26. Magnetic reconnection — 32
*Image reprinted by permission from Macmillan Publishers Ltd: P. Hanlon et al., **Nature Physics** (Aug. 1st, 2005), doi:10:1038/nphys111; copyright 2005*

Figure 27. Solar flare — 33
Image courtesy of SOHO

Figure 28. Plots of solar wind speed — 34
*Image reprinted by permission from D. J. McComas. This image was published by AGU: D. J. McComas et al., **Geophysical Research Letters** 30 (2003), no. 10, p. 1517, doi: 10.1029/2003 GLO 017136; copyright 2003*

Figure 29. Interaction of Venus with solar wind — 35
Image courtesy of European Space Agency

Figure 30. Earth's magnetosphere — 36
Image courtesy of NASA /Marshall Space Flight Center

Figure 31. Jupiter composite with aurorae — 37
X-ray image courtesy of NASA/CXC/SWRI/R. Gladstone et al.; optical image courtesy of NASA/ESA/Hubble Heritage (AURA/STScI)

Figure 32. Eugene N. Parker — 38
Image courtesy of Eugene N. Parker

Figure 33. Fission vs. fusion — 39

Figure 34. D–T fusion — 40
Image revised from Physics 2010 Committee, **Plasma Science: Advancing Knowledge in the National Interest** *(Washington, D. C.: The National Academies Press, 2007), p.19; with permission from ITER*

Figure 35. The tokamak — 41
Image revised from Physics 2010 Committee, **Plasma Science: Advancing Knowledge in the National Interest** *(Washington, D. C.: The National Academies Press, 2007), p. 20; with permission from ITER*

Figure 36. International Thermonuclear Experimental Reactor — 42
Image courtesy of ITER

Figure 37. Plasma heating in a tokamak — 43
Image copyright UKAEA, Culham Science Centre

Figure 38. Computer simulations of magnetic islands in a tokamak — 44
Simulation by Scott Kruger using the NIMROD code on flagship computers at the National Energy Research Scientific Computing Center (NERSC). Visualization and analysis by Allen Sanderson using the SCIRun Problem Solving Environment. Support provided by the US Dept of Energy Scientific Discovery Through Advanced Computing (SciDAC) Project

Figure 39. Computer simulations of plasma turbulence — 45
Image courtesy of Zhihong Lin, University of California, Irvine, Department of Physics and Astronomy

Figure 40. Direct vs. indirect drive in inertial confinement — 46
Image revised from Physics 2010 Committee, **Plasma Science: Advancing Knowledge in the National Interest** *(Washington, D. C.: The National Academies Press, 2007), p. 86*

Figure 41. Hohlraum — 47
Image courtesy of Lawrence Livermore National Laboratory, the National Ignition Facility, and the Department of Energy, under whose auspices the work was performed

Figure 42. Computer simulation of density irregularities in inertial confinement target — 47
Image courtesy of Lawrence Livermore National Laboratory, the National Ignition Facility

Figure 43. Laser bay of National Ignition Facility — 48
Image courtesy of Lawrence Livermore National Laboratory, the National Ignition Facility

Figure 44. Marshall Rosenbluth — 50
Image courtesy of Laura Moore, University of California, San Diego

Figure 45. "The Arches" — 51
X-ray image (blue) courtesy of NASA/CXC/SwRI/R.Gladstone et al.; optical image (red) courtesy of NRAO/AUI/NSF

Figure 46. 3-D computer simulation of the geodynamo — 53
Image courtesy of Gary A. Glatzmaier, University of California, Santa Cruz, and Paul H. Roberts, University of California, Los Angeles

Figure 47. Neutron star at the core of the crab nebula — 54
Image courtesy of NASA/CXC/ASU/J. Hester et al.

Figure 48. Massive star, gamma-ray burst, supernova, magnetar — 55
Images courtesy of NASA E/PO, Sonoma State University, Aurore Simonnet

Figure 49. Core of galaxy NGC 4261 — 57
Image courtesy of National Radio Astronomy Observatory, California Institute of Technology; Walter Jaffe/ Leiden Observatory, Holland Ford/ JHU/STScI, and NASA

Figure 50. Computer simulation of density variations in matter surrounding a black hole accretion disc — 58
Image courtesy of J. Stone, Princeton University

Figure 51. Jet from M87 galaxy — 59
Image courtesy of NASA

Figure 52. Fluxes of cosmic rays — 61
Image courtesy of Simon Swordy, University of Chicago

Figure 53. Astrophysical plasma cloud — 62
Image courtesy of Philipp P. Kronberg et al., Los Alamos National Laboratory / Arecibo Observatory / DRAO. Reprinted with permission of the AAS: **The Astrophysical Journal 659** *(2007), p. 269; copyright 2007*

Figure 54. Putting plasmas to work — 63
Images courtesy of www.dreamstime.com

Figure 55. Schematic of fluorescent light — 64
Image courtesy of David P. Stern

Figure 56. Fluorescent tube — 65
Image courtesy of Paul Kevin Picone/PI Corp and OSRAM SYLVANIA Inc.

Figure 57. United States from space at night — 65
Image courtesy of Craig Mayhew and Robert Simmon, NASA-GSFC, based on Defense Meteorological Satellite Program data

Figure 58. Plasma TV in the living room — 66
Image courtesy of www.dreamstime.com

Figure 59. Building a microchip — 67
Image courtesy of Jeffrey Hopwood, **About Plasmas – Computer Chips and Plasmas** *(2006); with permission from the Coalition for Plasma Science*

Figure 60. Carbon cathode arc used in coating carbon plasma — 68
Image courtesy of Blake Wood, Los Alamos National Laboratory

Figure 61. Low-temperature plasmas synthesizing silicon nanoparticles — 69
Image courtesy of A. Bapat et al. and the American Institute of Applied Physics

Figure 62. Plasma treatment on *E. Coli* microbes — 70
Image courtesy of M. Laroussi et al. **Applied Physics Letters 81** *(2002), no. 4, pp. 772–774; copyright 2002*

Figure 63. Plasma-based ion thrusters — 71
Image courtesy of Yevgeny Raitses and Nathaniel Fisch, Princeton Plasma Physics Laboratory, Hall Thruster Experiment

Preface

On the occasion of the 50th Anniversary of the Division of Plasma Physics (DPP) of the American Physical Society (APS), the Executive Committee of the DPP approved a plan to publish a book on plasma physics for the popular audience. We imagined a book that told the story of plasma physics through both text and visual images. The story is worth telling. In the last 50 years, the field has grown and matured in several directions, with very broad impact. The pursuit of controlled thermonuclear fusion by magnetic and inertial confinement has progressed to the point where the multinational International Thermonuclear Experimental Reactor in France and the National Ignition Facility in the USA is expected to demonstrate self-sustaining fusion reactions for the first time. Application of the fundamental principles of plasma physics has provided models for a rich variety of phenomena in solar–terrestrial plasmas, and is providing crucial insights into longstanding mysteries of the cosmos. Tabletop accelerators based on laser–plasma interactions have demonstrated the potential to achieve remarkable particle acceleration over short distances, and are among the most promising ideas for future particle accelerators that will probe the fundamental structure of matter. And plasmas have been put to work in a variety of ways that have arguably enhanced the quality of our lives: producing the lights that illuminate our living spaces and the roads we travel, the striking plasma television sets that bring the world into our living rooms, and the manufacturing of chips that constitute many of the electronic devices we rely on every day. All of this has become possible by the interplay of ingenious experiments with sophisticated mathematical models and computer simulations that plasma physicists have excelled at, producing many of the beautiful images that fill this book. Viewed in its entirety, this is a fabulous story, and the whole is much greater than the sum of its parts.

To tell this story, we commissioned Curt Suplee, prize-winning science writer and author of *Physics in the 20th Century*, which was published on the occasion of the APS Centennial. Kristi Donahue, graphics designer at the Institute for the Study of Earth, Oceans, and Space at the University of New Hampshire masterfully integrated Suplee's text with images. My contribution to this collaboration included the selection of topics, sources, and the images. I tried to stay out of the way of my creative collaborators, except to make the narrative as technically accurate as my own knowledge of plasma physics allowed. Since the book covers many more topics than I have expertise in, I turned for help to an astute advisory committee, which provided valuable assistance. Ian Hutchinson deserves special mention for reading every chapter and offering detailed advice. I would also like to acknowledge Vincent Chan, Richard Hazeltine, Chandrasekhar Joshi, Rick Lee, Michael Mauel, and Fred Skiff for their encouragement and advice. Provost Bruce Mallory of the University of New Hampshire provided a generous subvention that supported a fraction of Kristi Donahue's time and allowed us to pay for some of the images included in this book. Last but not the least, I would like to thank my two-year old son Arun (whose name literally means the charioteer of the Sun) for giving up some of his play time with me so that I could work on this book. In another 50 years, on the occasion of the DPP Centennial, it is my hope that he will bear witness to the great promise of plasma physics reflected in the pages of this book.

Amitava Bhattacharjee
Chair, Division of Plasma Physics, The American Physical Society

The Fourth State of Matter

LIKE "WEATHER," *plasma* is a deceptively simple word signifying a complicated, fascinating and baffling range of phenomena. No succinct definition quite contains its dazzling variety.

A plasma is, at its most fundamental level, an "ionized gas." But that phrase doesn't really capture the cozy glow of neon lights that beckon nocturnal strollers on city streets, or the gossamer shimmer of the aurora borealis. And it certainly doesn't convey the violence with which the Sun's plasmas routinely blast hundreds of millions of tons of mass out into space at a million miles an hour.

Alternatively, a plasma can be defined as a collection of charged particles whose dynamics are chiefly governed by electromagnetic forces. But that seems a bit tame to describe what goes on in nuclear fusion reactors, where the temperature hits 100 million degrees in a seething mayhem of particles trapped by invisible magnetic fetters. Nor is it adequate to depict the whirling inferno of plasma that forms around a black hole as it sucks whole stars into oblivion.

Even the most popular descriptions miss the mark. Plasmas are often called the "fourth state of matter," to distinguish them from the more familiar solids, liquids and gases that fill our world. That's a useful distinction. But it's only meaningful in a few odd spots in the universe where solids, liquids and gases can actually exist. One of those rare locations is our pleasant little planet, an island of tepid energies in which nearly all the atoms we encounter are unbroken, well-behaved and electrically neutral. Each of these humdrum, "normal" atoms contains exactly the same number of positive charges – from the protons in its nucleus – and negative charges from its orbiting electrons. So here on Earth, with a few spectacular exceptions, such as lightning and the high auroras, plasmas almost never arise naturally, although we create them every time we light a burner on a gas stove, flip on a fluorescent light, fire up an arc welder or sit down to watch a plasma TV screen.

But in most of the universe, plasma is actually the "first" state of matter, and by a very large margin. Out there, fully intact atoms are abnormal in the extreme. Practically all the visible contents of the cosmos – not just stars, but even regions of rarefied interstellar dust containing barely a million particles per cubic meter – are in a plasma state. That is, they consist of positively charged ions (atoms that are missing one or more electrons) and an associated population of negatively charged free electrons that may once have belonged to those atoms but were ripped out of their orbits by irresistible energies.

The ubiquitous persistence of plasmas may at first seem counterintuitive. After all, since positive and negative charges attract one another – and since nature famously favors equilibrium, the lowest-energy state in any physical system – then why don't the ions and electrons just recombine? Well, sometimes they do, if the temperature is below about 1000 K.[1] Above that, the energy involved is simply too fierce.

How fierce? The energies required to create plasmas can span many orders of magnitude – all of them far from the parameters of everyday existence. For example, consider the simplest and most common atom, hydrogen. Its single electron is tightly bound to its single proton, and it takes an energy of 13.6 electron volts to overcome that attachment and separate the electron from the nucleus. That might sound like a small amount, since one electron volt (1 eV) is defined as the amount of energy that one individual electron gains when it moves through a potential of one volt in a vacuum. But 1 eV is equivalent to a temperature of about 11 600 K: about four times as hot as the filament in an incandescent light bulb and twice the temperature on the surface of the Sun. And that's downright chilly compared to a fusion reactor, where energies can reach over 10 000 eV.

Whatever the energy, plasmas occur on a dizzying range of spatial scales. In a plasma TV screen, each pixel is around 0.3 mm long. The industrial plasmas that help create the microchips in that TV are a few centimeters wide. A fluorescent tube in a ceiling fixture is around one meter in length. The tenuous plasma that begins to form about 70 km above the Earth's surface – called the ionosphere, and produced when high-energy solar radiation hits the outermost layers of the atmosphere – is hundreds of kilometers thick. The solar wind, a torrent of charged particles that blows off the Sun in all directions, extends for hundreds of millions of kilometers. And the frigid gas plasmas between the stars stretch for tens of trillions of kilometers.

IN A PLASMA TV SCREEN, EACH PIXEL IS AROUND 0.3 mm LONG.

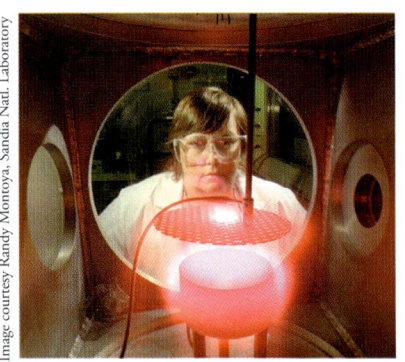

THE INDUSTRIAL PLASMAS THAT HELP CREATE THE MICROCHIPS IN THAT TV ARE A FEW CENTIMETERS WIDE.

---

[1] *Physicists customarily express temperature in units of kelvin (named for the nineteenth-century scientist and abbreviated K), a practice followed here. One kelvin is the same magnitude as a Celsius degree. But, instead of setting the zero point at the temperature at which water freezes at sea level, the kelvin scale begins at absolute zero – about −273 °C. So a pleasant room temperature of around 70 °F, or 21 °C, is 294 K.*

A FLUORESCENT TUBE IN A CEILING FIXTURE IS AROUND ONE METER IN LENGTH.

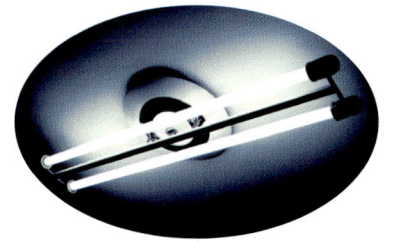

The tenuous plasma that begins to form about 70 km above the Earth's surface – called the ionosphere, and produced when high-energy solar radiation hits the outermost layers of the atmosphere – is hundreds of kilometers thick.

Plasmas produce one of Earth's paramount special effects: the aurorae that form when particles from the solar wind are trapped in, and excited by, the planet's magnetic field lines. As the plasma particles shed their newly gained energy, they emit photons – many of them in visible wavelengths.

EXAMPLES OF PLASMAS

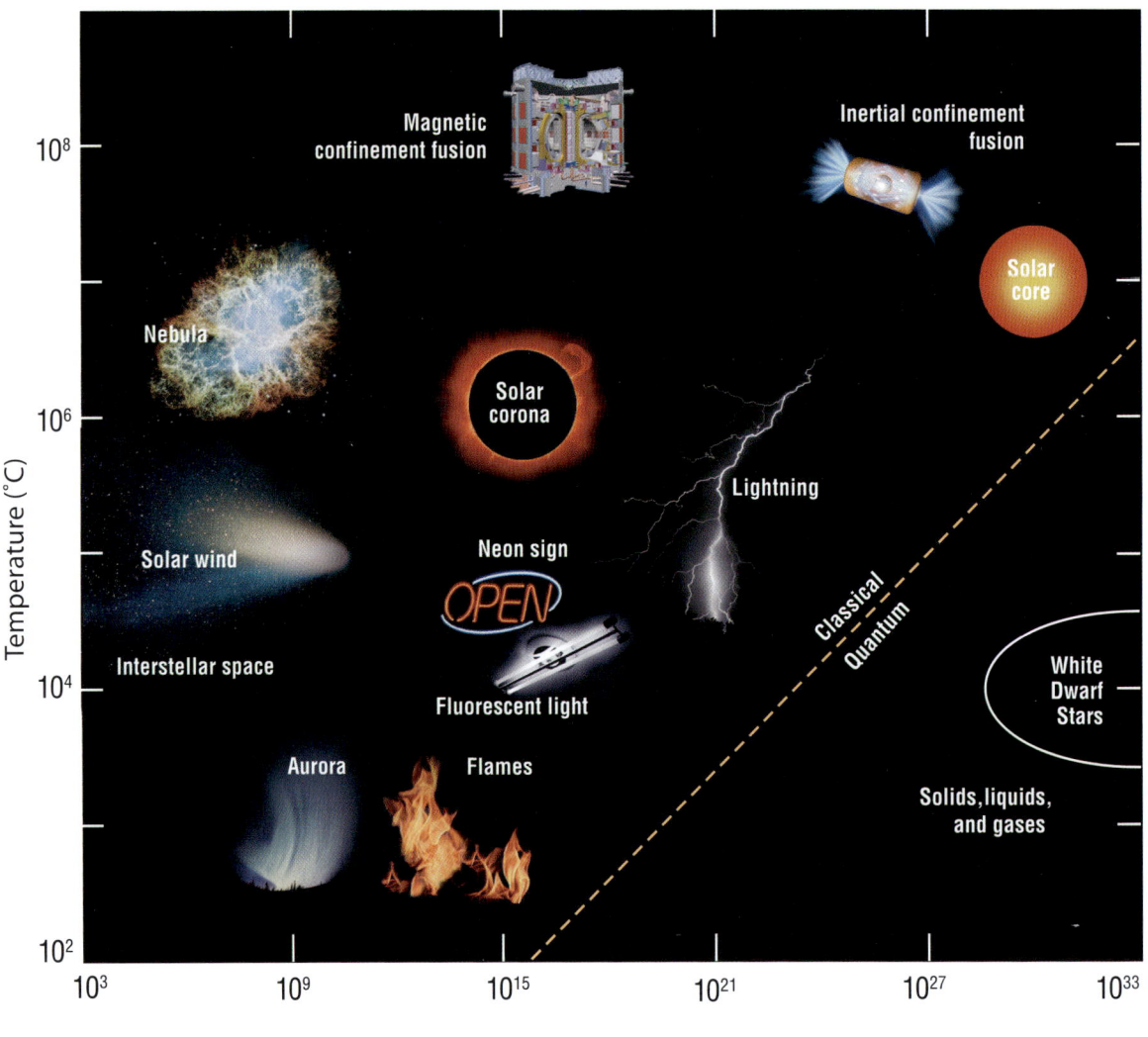

Plasmas form in an extraordinary range of conditions, from the rarefied dust between the stars to the inferno of nuclear fusion. Those above and to the left of the dashed line comprise the classical plasmas. On the other side are those often studied for their quantum-mechanical properties.

Magnetic confinement fusion image courtesy of ITER; nebula, solar wind and solar corona images courtesy of NASA; inertial confinement image courtesy of the Lawrence Livermore National Laboratory, the National Ignition Facility, and the Department of Energy, under whose auspices the work was performed.

The Same But Different

> HOW CAN ONE BRANCH OF SCIENCE possibly encompass events that take place over such a staggering span of energy levels, physical conditions and spatial dimensions? The answer lies in one of many remarkable characteristics of plasmas: they all share numerous common properties irrespective of scale – properties that make them behave very differently from the kinds of matter we encounter in our daily lives, and that allow researchers to study and model plasma dynamics from a few atoms to astronomical formations.

Nearly all of this understanding has been gained within the past 100 years, and it arrived in halting increments. Indeed, there was no suitable theory explaining the role of ions (from the Greek for "things that move") in physics or chemistry until the late nineteenth century. Experimenters knew that water, which won't conduct electricity in its pure state, became a splendid conductor when salts were dissolved in it. But even the great British experimentalist Michael Faraday (1791–1867), who coined the term "ion," didn't know why. Swedish chemist Svante Arrhenius (1859–1927) proposed a then-radical explanation. The sodium and chlorine atoms that make up ordinary table salt (sodium chloride) split apart in solution into two electrically charged ions – positively charged sodium ions and negatively charged chlorine. The presence of those free charges allows current to pass easily through the liquid.

More than a century later, we know that a somewhat analogous process is at work to enable lightning bolts to travel through air, which, like nearly all other gases, is a notorious non-conductor. During a storm, powerful convection currents force large moving volumes of air and water droplets to rush past one another. As they do so, huge electrical charges build up in the surrounding clouds through the same kind of contact process well known to schoolchildren from classroom demonstrations in which a glass rod is rubbed by a piece of silk. Eventually, the pent-up charges reach a point at which a trickle of current starts to move through the atmosphere. The energy of that current ionizes a portion of the air molecules, changing them from world-class insulators into a conductive plasma, and forming a path that carries the bolt we see as lightning.

Of course, Arrhenius conceived his theory before the structure of the atom was known – indeed, well before the modern concept of the atom was fully accepted. It was not until 1897 that British physicist J. J. Thomson (1856–1940) identified the eerie "cathode rays" that streamed off heated electrodes mounted in vacuum tubes. Contrary to expectation, they were not radiation at all, Thomson determined. Instead, they were streams of tiny, extremely low-mass particles that eventually became known as electrons. That and similar discoveries, along with a burgeoning interest in the propagation and control of radiation and electrical currents in vacuum and rarefied gases, prepared the way for the confluence of diverse research interests that – in the course of more than half a century – formed the modern science of plasma physics.

In fact, some activity was well underway before the field had a name: the use of the word "plasma" (from the Greek for something molded) in physics is a fairly late linguistic arrival. American physical chemist Irving Langmuir (1881–1957) coined the usage during his extensive study of ionized gases. He theorized that some entity underlay and shaped the assemblies of charged particles that displayed strikingly coordinated behavior. He called this mystery substance "plasma" by analogy with the fluid that surrounds and transports different kinds of blood cells.

> "Except near the electrodes," Langmuir wrote in 1928, "where there are sheaths containing very few electrons, the ionized gas contains ions and electrons in about equal numbers so that the resultant space charge is very small. We shall use the name plasma to describe this region containing balanced charges of ions and electrons."

Consequently, Langmuir is regarded as the principal founder of the field. However, at least five quite different lines of inquiry shaped the early evolution of plasma physics.

RADIO

In 1901, Guglielmo Marconi's famous communication system – regarded by the general public as a modern marvel – became amazing even to experts. That year, Marconi (1874–1937) succeeded in sending radio waves across the Atlantic Ocean, demonstrably contrary to scientific consensus, which held that electromagnetic radiation must travel in straight lines and could not possibly move around the curve of the Earth. What was going on? In 1902, scientists Oliver Heaviside (1850–1925) in England and Arthur Kennelly

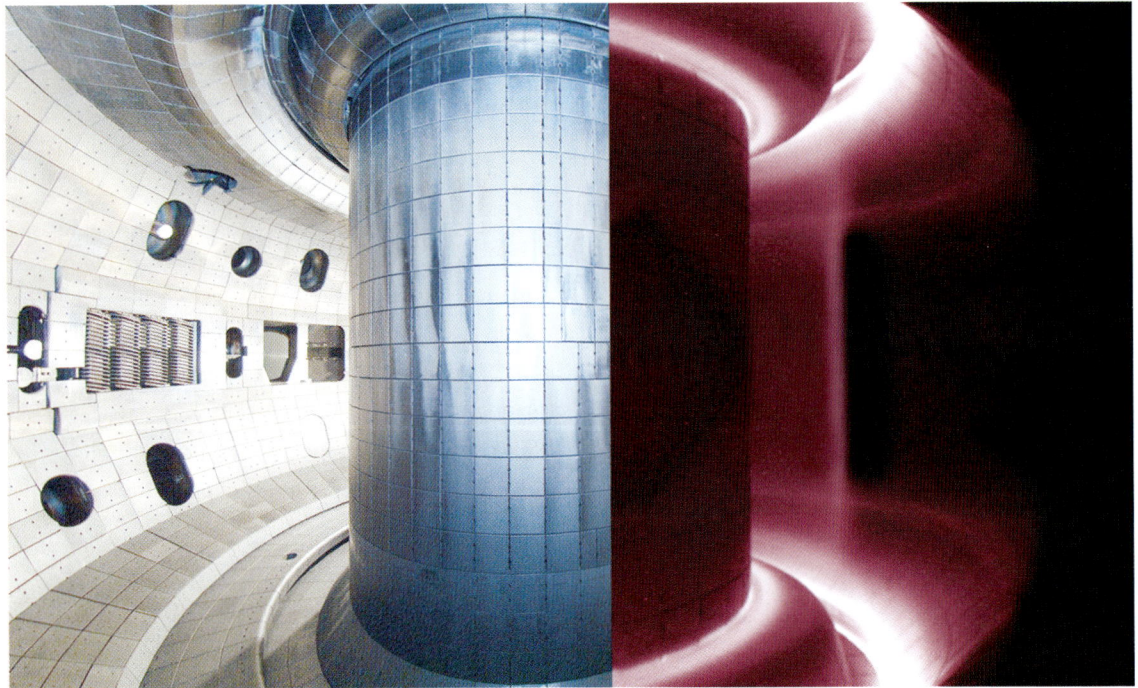

The DIII-D magnetic-confinement fusion device, maintained by General Atomics near San Diego, CA, has a plasma torus chamber about 9 feet high. The image at right shows the enclosure with plasma in place.
Image courtesy of General Atomics.

(1861–1939) in the United States independently hypothesized the same explanation: there must be some ionized layer of the atmosphere that reflected the radio waves like a flashlight beam bouncing off a mirror. It took two decades until English physicist Edward Appleton (1892–1965) employed a British Broadcasting Corporation transmitter to test the hypothesis, confirming the existence of what is now called the ionosphere in 1924. Appleton and others went on to develop a systematic theory of how magnetic waves propagate in a non-uniform plasma, thereby making it possible to comprehend, and hence correct, distortions in radio waves.

ASTROPHYSICS

By the 1930s, it had become clear that many astrophysical events resulted from plasma activity. But their mechanisms were poorly understood until Swedish physicist Hannes Alfvén (1908–1995) provided a profound and widely applicable theoretical basis in the form of magnetohydrodynamics – the rules governing the behavior of electrically conducting fluids (see Chapter 2), including the generation of planet- and star-size magnetic fields. Today, plasma science is integral to a host of astronomical investigations, including an urgent worldwide effort to understand and predict the nature of solar storms and mass ejections – the projectile plasmas from the Sun which pose a lethal threat to the satellite-based global communications systems and even electrical power facilities on the Earth's surface.

WEAPONS AND MAGNETIC FUSION

Shortly after World War II, which ended with an epochal demonstration of nuclear fission, scientists finally found a way to unlock the even greater power of nuclear fusion, whereby lightweight elements such as hydrogen combine into heavier elements such as helium, shedding energy in the process. It made its most dramatic appearance in military tests of "H-bombs." But after fusion research was generally declassified in 1958, scientists began publishing influential papers that fundamentally changed the nature of plasma physics, giving it a highly rigorous mathematical basis to accompany the accumulated wealth of experimental data and observations. At the same time, physicists and engineers began work on ways to harness nuclear fusion for civilian power generation (see Chapter 4), a global initiative that continues to this day.

GEOSCIENCE AND THE MAGNETOSPHERE

Yet another line of research that emerged from plasma studies changed our understanding of the Earth's immediate surroundings. An American space scientist, James Van Allen (1914–2006), used leftover German V-2 rockets and his own configuration of balloon-borne rockets ("rockoons") to probe ever higher in the far upper atmosphere to study cosmic rays and related phenomena. In 1958, one of his experiments – mounted on America's first space satellite, Explorer – returned historic evidence of regions containing energetic charged particles trapped by the Earth's magnetosphere. These formations, now called the Van Allen radiation belts, along with associated elements of the near-Earth environment, became the objects of an intense research effort: the new field of space plasma physics. Half a century later, it is still producing illuminating insights of both scientific and practical value.

LASERS AND INERTIAL FUSION

Finally, the invention of the laser in 1960 (based on a prediction about photon behavior made by Einstein decades earlier), and the advent of its progressively more powerful successors, ushered in an entirely new way of generating and studying plasmas. The impressive energy contained in a stream of coherent photons, each perfectly in phase with its neighbors, can vaporize many objects, producing a plasma in the space between the beam and the target. The resulting plasmas have uniquely interesting and unconventional properties, such as densities approximating those of solids. In recent years, high-tech lasers have proven capable of turning plasmas into desktop particle accelerators (see Chapter 2), and of generating a barrage of photons so energetic that they force tiny pellets of hydrogen isotopes to fuse into helium (see Chapter 4).

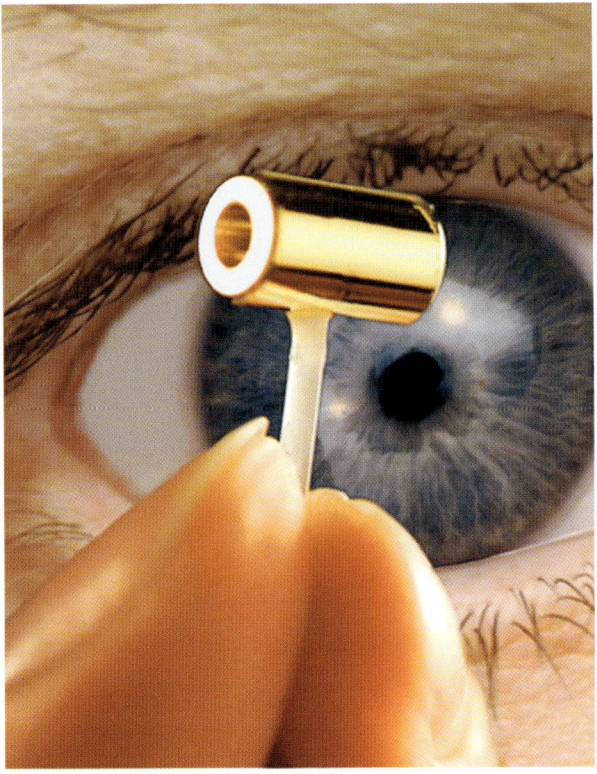

A golden "hohlraum" used for laser fusion. The hohlraum is a container designed to heat the fusion fuel uniformly to a very high temperature.

Image courtesy of Lawrence Livermore National Laboratory, the National Ignition Facility, and the Department of Energy, under whose auspices the work was performed.

On facing page: a non-neutral plasma composed of electrons can behave like a conventional fluid. The images are end views of a cylindrical electron plasma. A circular electron vortex (first image from left) is deformed into an elliptical patch by applying an oscillatory electric field (second image). The elliptical vortex evolves spontaneously by rotating and producing extended finger-like filaments (third image) that eventually break off, leaving a new elliptical vortex (fourth image).

Image courtesy of Travis B. Mitchell; research conducted at the Department of Physics and Astronomy, University of Delaware, under the auspices of the Department of Energy and the National Science Foundation.

What's In a Name?

THE BROAD SCOPE OF SUBJECTS covered in the preceding brief history may suggest that the word "plasma" can apply to virtually any collection of ionized particles. But that is not the case, and physicists have restricted the term to apply only to gases that meet certain requirements. One is Langmuir's concept of "balanced charges." In most cases, the number of positive and negative charges in the ensemble of particles must be approximately equal, with the result that the plasma as a whole is neutral or, more accurately, "quasi-neutral." There may be local regions of a small net negative or positive charge, just as certain zip codes in a state may have more women than men, or vice versa. But throughout large volumes, the number of electrons and ions is nearly the same. Of course, some plasmas may also contain neutral atoms or molecules in the mix, but that does not change the charge-balance requirement.

(There is a maverick subcategory of plasmas that does not obey that rule. These entities, appropriately called "non-neutral," are mostly or entirely made up of particles with a single charge: electrons, ions, protons or – in their most exotic form – anti-electrons or anti-protons. Unlike their neutral cousins, they can be firmly confined by electrical and magnetic fields indefinitely for study. Non-neutral plasmas display some characteristics of quasi-neutral plasmas, but they also demonstrate novel behaviors, such, as the spontaneous formation of geometrical arrays like the atoms in a crystal.)

Today, the standard definition of a plasma also includes a second criterion, which involves the natural "shielding" effects that occur in populations of charged particles. Imagine a neutral plasma, with an equal distribution of charges throughout a given volume, into which an object with a strong positive charge is suddenly introduced. Immediately, electrons will begin to drift toward the intruder and adjacent ions will be pushed away slightly. As the electrons gather, their abundance creates a region of net negative charge; as the ions recede, their positive charges become more concentrated. Where there was once an equal average distance between ions and electrons, there is now clustering that produces a polarity.

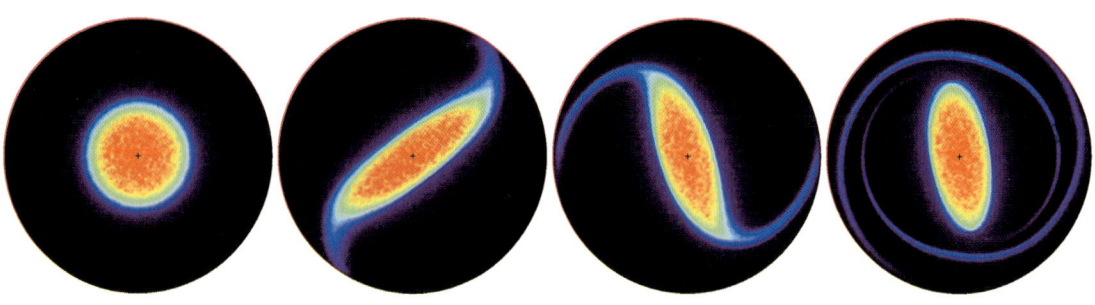

Irving Langmuir
(1881–1957)

PHYSICAL CHEMIST

NOBEL PRIZE, 1932,
for "discoveries and investigations in surface chemistry."

Irving Langmuir

THE MAN WHO GAVE A NAME to the fourth state of matter was equally at home in theory, experiment, invention, and manufacture. Irving Langmuir was born in Brooklyn, NY, to parents who encouraged close observation of nature, as did his older brother, a research chemist. Their efforts succeeded, as he graduated from Columbia University's School of Mines and received his doctorate in 1906 from Gottingen, Germany, where his work chiefly involved effects of energetic discharges into rarefied gases.

By 1909 he had joined General Electric's celebrated research laboratory in Schenectady, NY, where he would remain for 41 years. Almost immediately he distinguished himself by inventing the diffusion pump for creating high vacuums, and by introducing inert gases (such as nitrogen and argon) into electric light bulbs to extend significantly the life of the filament.

Langmuir's originality and multi-faceted genius ranged into theoretical chemistry (in which he extended the ideas of Gilbert Lewis about electron orbits and chemical bonds) and surface chemistry, where he advanced the understanding of catalysis. He also discovered a form of wind-driven ocean circulation, formulated a popular system of cloud-seeding, and identified many reactions and processes that occur in liquid–gas interfaces.

And, of course, he conducted the first highly systematic investigations of plasma phenomena, during which he invented the diagnostic electrostatic probe, still used in plasma physics today, and a hydrogen blowtorch that produced the first "plasma welding."

Langmuir was an uncommonly careful experimentalist, and is well-known for naming and illustrating a category of research error he called "pathological science," namely: "…cases where there is no dishonesty involved but where people are tricked into false results by a lack of understanding about what human beings can do to themselves in the way of being led astray by subjective effects, wishful thinking or threshold interactions."

But this polarity does not spread through the entire plasma. In fact, its very existence acts as a shield, which prevents particles at a distance from "feeling" the intruding positive object. This effect is easy to picture with a comparable social situation. Suppose there is a large cocktail party made up of celebrities and journalists. After half an hour or so, the distribution of each type will average out across the room. If a new celebrity abruptly arrives, nearby journalists will move toward the newcomer – obscuring the view of the new arrival and causing a temporary surplus of unaccompanied celebrities in a depleted circle around the event. But soon there is no more space for journalists around the new celebrity, and those on the fringes drift back toward the rest of the party, resuming conversations. After this shuffling, there will be a disproportionately dense gaggle of journalists around the arrival and a slight surplus of celebrities in the surrounding vicinity. But guests within 15 or 20 feet of the new arrival will have no idea that she showed up. As far as they can see, nothing new has happened. Of course, the "no recognition" distance will vary according to the total number of guests and the speed with which the journalists move among the celebrities.

In plasmas, the distance over which the shielding effect takes place is called the Debye length, after Dutch–American physicist Peter Debye (1884–1966) who first studied the process in solutions. Because plasmas are three-dimensional, the effect ordinarily takes the shape of a sphere that is larger or smaller depending on the density of the particle concentration and the temperature of the particles. An ionized gas is deemed a plasma if its dimensions are substantially greater than the Debye length.

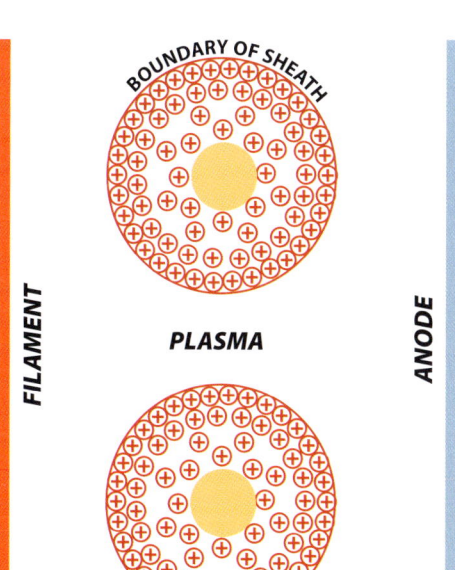

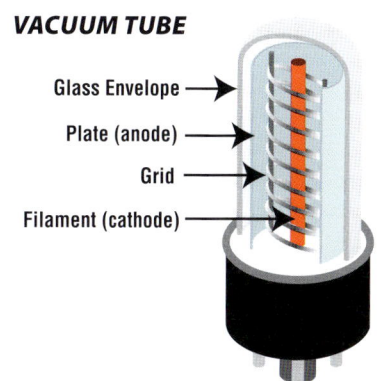

In a celebrated paper published in 1929, Irving Langmuir and colleague Albert Hull described the formation of "plasma sheaths" (red positively charged ions) around negatively charged grid wires (gold) in a vacuum tube (left frame). The function of the grid is to control the flow of current from the hot filament (cathode) to the anode, and the researchers explained how the plasma effects contribute to the tube dynamics. A typical vacuum tube is shown above.

Left illustration drawn after Albert W. Hull and Irving Langmuir, *Proceedings of the National Academy of Sciences, USA* (March 15, 1929).
Right illustration modified from triode tube schematic, licensed under Creative Commons Attribution-Share Alike 2.5 Netherlands.

Forces and Feedbacks

ALTHOUGH QUASI-NEUTRALITY and Debye shielding are the fundamental discriminating characteristics of plasmas, those qualities don't begin to account for the myriad effects and protean forms that are possible. The combined and interactive effects of electrical and magnetic fields give plasmas their seemingly magical repertoire of variety and complexity.

As the lightning-bolt example demonstrates, it takes shockingly little ionization to turn an array of stolid neutral atoms into a gas with lower electrical resistance than pure copper wire. (Unlike metals – which obey Ohm's law and generally become less conductive the hotter they are – the electrical resistance of plasmas decreases with the amount of current they carry.) Although some fraction of ions and electrons will always recombine into neutral atoms, ionization is a rapidly escalating activity because each newly dismembered ion and electron speed off into collisions with surrounding atoms, creating new ionizations. And the thresholds at which resistance drops are low. In fact, a gas can reach 50 percent of its maximum possible conductivity when as few as one out of every 1000 atoms is ionized. And it typically takes only one in 100 to achieve the highest conductivity.

Along the way, the convoluted interplay of electrical and magnetic forces begins to produce effects of astonishing complication. Any moving charge – including the charged particles in a plasma – creates a magnetic field, and with it the field lines along which charged particles are constrained to move. But those moving magnetic fields induce currents to flow, and those currents in turn generate new magnetic fields, and so forth. In every plasma, both effects are constantly in action, each affecting the other in a tangle of mutually reinforcing feedbacks.

This situation is made yet more complex because electrons and ions respond to changing plasma conditions somewhat differently. For one thing, they have a whopping disparity in mass. An electron has only about 1/2000 the mass (and hence the inertia) of a proton. That is, for the same amount of force, it is about 2000 times easier to move an electron than it is to move a proton. It is, therefore, not surprising that electron and ion motion can often be decoupled. This leads to distinctively peculiar outcomes.

Given all these perplexing processes, a casual observer might reasonably conclude that plasmas are hopelessly incomprehensible, and turn to more tractable problems such as ending world hunger. Fortunately for science and industry, this is by no means the case, as succeeding chapters will demonstrate. Researchers have come to understand many of the intricate riddles of plasmas on multiple scales. In so doing, they have also discovered that a plasma is not simply the integrated sum of each individual particle's effects on its immediate neighbor. The constituents of plasmas respond to long-range forces, which give rise to elaborately choreographed collective motions and wave-like behaviors that can be described, modeled and utilized.

As the lightning-bolt example demonstrates, it takes shockingly little ionization to turn an array of stolid neutral atoms into a gas with lower electrical resistance than pure copper wire.

When huge voltages build up between clouds and the ground, a current starts to bridge the gap, ionizing air molecules and providing a conductive pathway for current discharges averaging around 40 000 amperes.

In 100 billionths of a second, the "Z-pinch" inertial confinement device at Sandia National Laboratory in New Mexico channels 18 million amperes of current through a cage of ultra-thin wires about the size of a spool of thread. The resulting imploding plasma produces energy that might be used in a fusion device.

Image courtesy of Randy Montoya, Sandia National Laboratory.

The Music and Dance of Plasmas

IMAGINE LOOKING DOWN upon a grand ballroom as the dancing reaches full flourish. Dozens of couples glide and spin around the floor, forming intricate collective configurations that change in rhythmic unison. Then imagine trying to understand those motions without being able to hear the music and without knowing the dance steps. That was the position scientists were in at the beginning of the twentieth century, before the systematic study of plasma waves began.

Since then, physicists have identified a bevy of wave and oscillation activities ranging in spatial scale from very short (but longer than the Debye length) to so long that the wave encompasses the entire plasma. They occur over a wide spectrum of frequencies, and their complexity can be amazing.

Combining only a few kinds of basic mechanical motions with the interplay of electrostatic and magnetic forces will prompt the constituents of plasmas to form elaborate, self-sustaining, and self-modifying patterns that can transport particles, energy, and momentum in myriad ways.

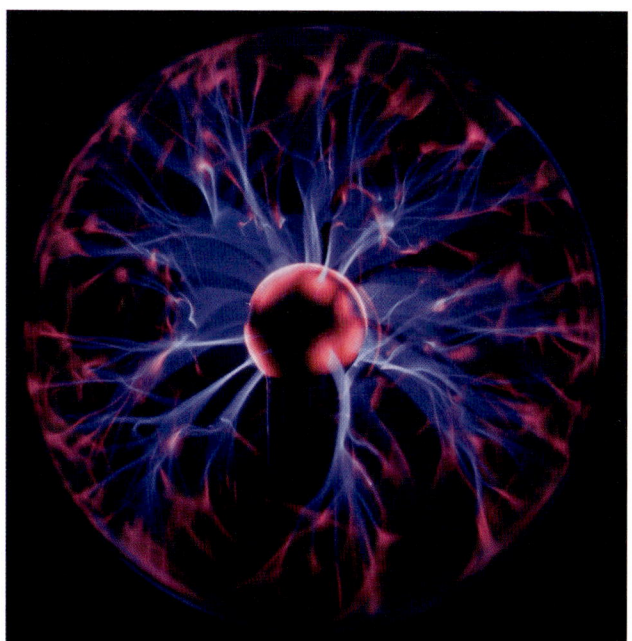

Image of a plasma lamp, defined in a Wiktionary (a wiki-based open content dictionary) as a clear glass orb, filled with a mixture of gases at low pressure, and driven by high-frequency alternating current at high voltage.

Image courtesy of Luc Viatour, licensed under the GNU Free Documentation License using material from Wikipedia.

Langmuir Waves (plasma oscillations)

PERHAPS THE MOST STRAIGHTFORWARD illustration of the transition from simple cause to complicated result occurs in the oscillations known as Langmuir waves, named after their discoverer. These waves involve only the motion of electrons in response to electrical forces, so the movement of ions can be neglected and magnetic fields do not play a role. And yet Langmuir waves produce spectacular effects – not only in stars, but also on the laboratory bench.

If a gas or neutral plasma is suddenly penetrated by a fast-moving beam of electrons or a strong laser pulse, the energy displaces electrons along the beam path, thrusting them outward. Because the beam's transit time through the gas is very brief, the energy transfer does not move the much heavier ions. They remain where they were, giving the electron-depleted region a net positive charge. Similarly, the dislocated electrons give their locations a negative charge. As the electrons start moving back in response to the charge imbalance, they gain momentum and end up overshooting their original positions until they are pulled back again by the electrical field, only to overshoot yet again. The cycle repeats in a back-and-forth motion called Langmuir waves.

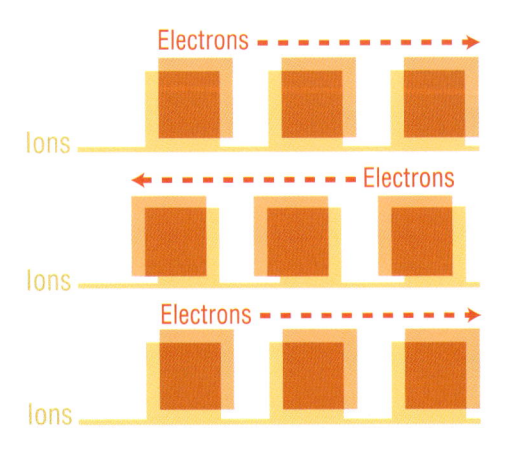

(The same sort of thing happens when a parent begins pushing a child on a swing. The child swings past the starting position, rising on the other side, only to fall backwards past the center point again. This sort of physical system is called a harmonic oscillator.)

When electrons are pushed around in this way, the frequency of the oscillations – also called the plasma frequency – is determined primarily by the number of electrons per unit volume. The greater the particle density, the greater the charge imbalance that provides the restoring force, and therefore the higher the frequency.

Taking that principle one step farther arrives at an intriguing conclusion: a beam passing through an increasingly dilute plasma should generate oscillations of progressively lower frequencies. And, as it turns out, that is exactly what was detected in a classic observation of Langmuir waves in the solar corona.

Langmuir waves (or plasma oscillations) are excited when electrons are separated from ions that are much heavier and remain nearly fixed (top frame). This separation of charges produces an electric field that tends to restore the electrons to their initial positions, but the velocity they acquire as they return makes them overshoot where they were (middle frame). The electric field changes sign, and electrons attempt to return, but overshoot again, setting up an oscillatory pattern of movement.

Adapted from F. Chen, *Introduction to Plasma Physics and Controlled Fusion* (New York: Plenum Press, 1984), p. 83.

Scientists had long known that certain kinds of solar flares were accompanied by emanations of electromagnetic radiation in the radio-frequency range that declined over time from hundreds of megahertz (MHz, millions of cycles per second) to around 10 kilohertz (kHz, thousands of cps). The mechanism that produces these "radio bursts" remained a matter of speculation, however, until the mid 1970s when the Helios spacecraft orbiting the Sun obtained data that confirmed the leading theory: the radio bursts are created when streams of electrons, expelled by solar flares, pass through the tenuous plasma called the solar corona that forms the Sun's outermost atmosphere, generating Langmuir waves as they go. Because the corona becomes less dense as it extends farther from the Sun, the electron streams provoke oscillations of lower frequency as they travel outward.

But Langmuir waves can also be studied in the closer and much more agreeable confines of a laboratory bench. And it is there that plasma researchers are making a major contribution to a related field – high-energy physics – while also extending the frontiers of their own.

TABLETOP ACCELERATORS

As lasers have become more powerful, they can now create extraordinary plasma effects in controlled experiments. One such phenomenon, called "wakefield acceleration," involves the production of ultra-fast matter waves by exploiting a handy byproduct of oscillations. Researchers shoot a pulse from a laser into a gas, and the energy transfer sets off Langmuir waves. As the laser pulse moves ahead, the electrons it displaces fall back toward their original location, much like the wake that forms behind a speedboat. The wake produces a strong negatively charged electrical field as the returning electrons cluster together before overshooting the center of the channel. When that happens, electrons injected into this region of high electric field are shoved in the direction of the departing laser pulse, accelerated by the field, and thus are able to "surf" in the wake.

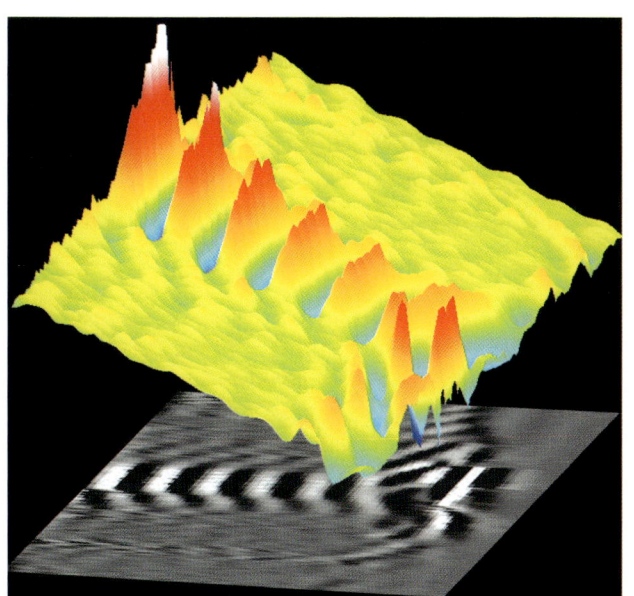

Physicists have photographed very fast Langmuir waves, moving at a speed of about 99.997 percent of the speed of light. These waves are generated in the wake of an extremely intense laser pulse in dense plasma, and give rise to enormous electric fields, reaching voltages higher than 100 billion eV per meter. The color image is a 3-D reconstruction of the oscillations, and the gray scale is 2-D projection of the data, showing curved wave crests.

Image courtesy of Nicholas Matlis and Michael Downer, University of Texas at Austin; Anatoly Maksimchuk and Victor Yanovsky, University of Michigan.

The result is a stream of electrons that travels at nearly the same pace as the laser pulse. In fact, the fastest matter waves ever seen – moving at an astonishing 99.997 percent of the speed of light – were produced when a series of pulses from a 30-trillion-watt laser punched through a helium plasma of about three billion particles per cubic centimeter, resulting in enormous voltages: about 100 billion eV per meter. By ingeniously manipulating two additional laser pulses, scientists have succeeded in taking a snapshot of the plasma oscillations.

Wakefield accelerators are also exciting researchers with their potential for a variety of practical benefits. They could become an inexpensive and compact source of focused streams of charged particles for use in cancer radiation treatments. And they could drastically lower the big cost of "Big Science" in fields such as particle physics.

For example, when researchers want to examine the most fundamental subatomic particles – such as the quarks and gluons that together make up a proton – or to search for particles predicted by theory but not yet seen, they typically turn to giant colliders such as the Tevatron at Fermilab outside Chicago, or the Stanford Linear Collider in California. These devices propel bunches of matter and antimatter into one another at approximately the speed of light by giving them repeated kicks from electromagnetic radiation as they race down the beam line. When the two bunches are steered into a collision, the energy released in the resulting annihilation congeals into matter via $E = mc^2$, Albert Einstein's illustrious equation describing the equivalence of mass and energy, in a spray of various exotic particles that are tracked and recorded by detectors that can be as large as an office building.

Not surprisingly, such colliders cost billions of dollars to build and operate. The world's most powerful, the Large Hadron Collider (LHC) outside Geneva, with a circular beam line 27 kilometers in circumference, could easily end up consuming as much as $10 billion before it realizes many of its science goals.

Might wakefield accelerators serve some of the same functions for a fraction of the price? The prospect is increasingly plausible. The LHC produces perfectly shaped bunches of protons with energies of 7 trillion eV per bunch. By contrast, the best wakefield accelerators have not succeeded in boosting particle energies beyond a few tens of billions of electron volts, and then only for a tiny percentage of the particles that are used to produce the wake in the first place. But they do so over a distance of less than a meter and at a cost that is vanishingly small compared with that of giant colliders. As research continues and techniques evolve, the energy density and beam quality are expected to improve dramatically, and many observers hope wakefield devices will soon have a major place in high-energy and medical physics.

Experimental Demonstration of Electron Acceleration by a Wake in a Meter-Scale Plasma

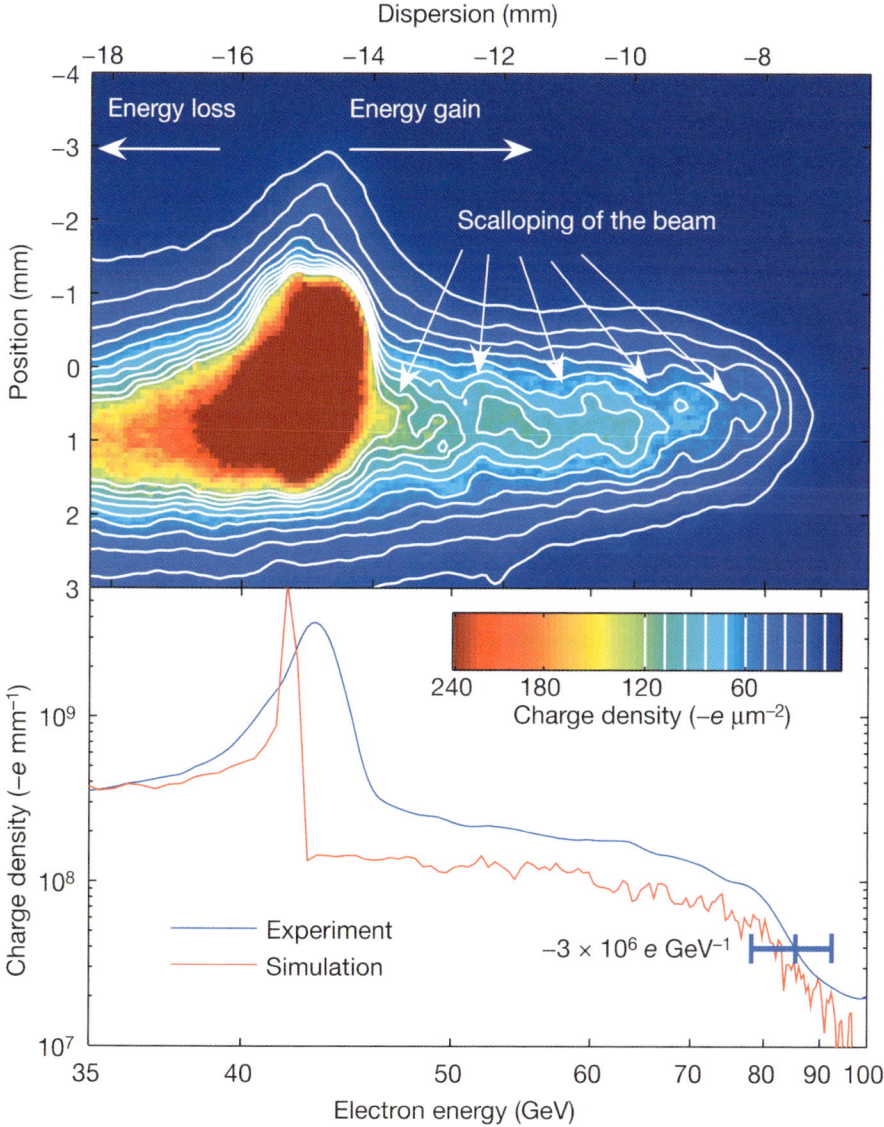

In this instance, the electrons in a pulse are accelerated by a plasma wake, which is driven by the pulse itself and then dispersed by a magnet that separates and "scallops" the beam electrons in space according to their energy (top frame). Initially the electrons in the pulse had a nominal energy of 42 GeV when they were injected into the plasma. Particles at the back of the pulse, which have reached energies up to 85 GeV, are visible to the right (bottom frame). The core of the electron pulse, which has lost energy driving the plasma wake, is dispersed out of the field of view of the camera. The energies indicated in red show the predictions of a computer simulation.

Image reprinted by permission from Macmillan Publishers Ltd: I. Blumenfeld et al., Nature 445 (2007), p. 742; copyright 2007.

Alfvén Waves Down the Line

ANOTHER FUNDAMENTAL TYPE of plasma wave at work in the depths of space, as well as nearby in the solar corona, solar wind, and the Earth's magnetosphere, involves very different kinds of interactions from those seen in Langmuir waves, including motion of ions and the presence of a strong magnetic field. This wave form is named for Swedish physicist Hannes Alfvén, who theorized its existence in the 1940s and received the Nobel Prize in Physics in 1970.

In general, charged particles are constrained to follow magnetic field lines, although they can have other kinds of motion as well, like raucous schoolchildren who are obliged to march in orderly rows on field trips. In Alfvén waves, both the ions and electrons of a plasma travel along magnetic field lines, oscillating at right angles to their direction of travel as they go. That motion is associated with an electromagnetic (EM) wave and involves oscillating electric and magnetic fields, unlike Langmuir waves, which are purely electrostatic and involve only oscillating electric fields. Because the ions are quite hefty compared with electrons, the characteristic frequency in Alfvén waves is significantly lower than that of Langmuir waves, often only a few cycles per second.

All EM waves have two components – an electric field and a magnetic field, perpendicular to each other – that change in exactly complementary ways over time.

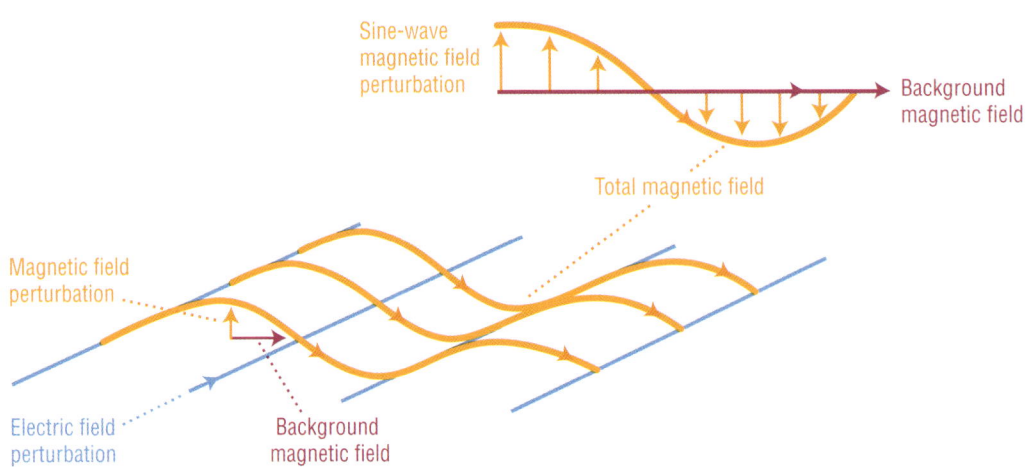

An Alfvén wave is produced when a background magnetic field, shown in dark red, is perturbed. The magnetic field perturbation causes a ripple (exaggerated in the figure above) and produces a perturbed electric field that, in turn, cooperates with the magnetic field perturbation to sustain the EM wave. The wave is analogous to the wave propagating on a string, with the background magnetic field playing the role of the tension in the string.

Adapted from F. Chen, *Introduction to Plasma Physics and Controlled Fusion* (New York: Plenum Press, 1984), p. 139.

In the case of Alfvén waves, the magnetic component of the oscillating plasma is perpendicular to the main magnetic field line; when the two are combined, the field line and particle ensemble develops a transverse motion. The effect is like plucking a guitar string with a motion at right angles to the string: the magnetic field line develops sine-wave-like ripples that propagate along its direction of travel. In fact, the whole system acts remarkably like a taut string that has beads strung along its length. In this case, the strength of the magnetic field corresponds to the tension on the string, and the oscillating plasma is equivalent to the string-and-bead mass per unit length. The field line and its attached ions move together as one object.

Alfvén waves were first observed under controlled conditions in 1959, and have been extensively studied since. But they are still making scientific headlines in the lab, thanks to sophisticated new facilities, high-tech instruments and ingenious experimental methods.

For example, in 2005, scientists at UCLA's Large Plasma Device (LAPD, a 68-foot-long tubular plasma chamber about 3 feet in diameter) were able to obtain unprecedented detailed, 3-D measurements of moving Alfvén waves. Inside the LAPD, the team created a cylindrical plasma of argon gas – 30 feet long and 16 inches wide, containing a few billion particles per cubic centimeter – and applied a magnetic field of 1500 gauss running straight down the plasma's long axis. An antenna at one end of the LAPD tube launched Alfvén waves at 1 Hz along the applied field line. A computer-controlled moving probe sampled the magnetic and electric fields of the waves at 1-foot spatial intervals in a huge series of 400-nanosecond "snapshots." The result was a collection of 8-by-16-inch cross-sectional data slices showing the dynamics of the moving waves at more than 15 000 different positions in the plasma.

Among other insights, this visualization technique has provided striking images of wave currents in plasmas of different elements and different excitation patterns. In a plasma of helium, Alfvén waves excited in a variety of vibrational modes revealed a seemingly haphazard tangle of currents. Remarkably, however, the overall current system was neither chaotic nor random: the same pattern occurred again and again over weeks of carefully reproduced experiments, and thus represents persistent physical principles. Finding mathematical descriptions for such newly available visualizations will be extremely valuable for plasma science in general and for other disciplines as well.

Of course, Alfvén waves are also studied in astronomical settings. They have now been identified as likely suspects in a longstanding mystery of solar physics – namely, why the corona is so anomalously hot. The visible "surface" of the Sun, called the photosphere, has a temperature of 5800 K. But the corona, which is a dozen orders of magnitude less dense and stretches millions of kilometers into empty space, has an average temperature around 2 million K. Why and how? No one is completely sure, but clearly some non-obvious mechanism is transferring energy from the star to its corona.

It was long suspected that, because solar magnetic field lines extend up into the corona and beyond, something associated with those lines must be

Alfvén waves propagating and interacting with each other in 3-D can produce a complex pattern of 3-D electric currents. In the figure above, obtained from a helium plasma in the laboratory, these currents have the appearance of interconnected snakes.

Image courtesy of Walter Gekelman, Department of Physics and Astronomy, University of California, Los Angeles.

heating the coronal plasma. Many scientists regarded Alfvén waves a probable candidate. Finally, in 2006, the Japanese solar orbiter Hinode ("Sunrise") began returning high-resolution images that unambiguously indicated the presence of Alfvén wave activity in the chromosphere –the layer between the solar surface and the corona. Although not definitive proof, the observations provide circumstantial evidence that heating occurs when the Alfvén waves encounter the coronal plasma, dissipating their energy. These observations also suggest that Alfvén waves contribute strongly to powering the solar wind.

Alfvén waves also show up in terrestrial fusion reactors. In magnetic-confinement devices, hydrogen plasma is tightly held in a toroidal (donut-shaped) configuration containing twisted magnetic fields that make it difficult for Alfvén waves to form. But in a few locations, the curvature of the magnetic field lines in a plasma is such that a few Alfvén frequencies are possible. At these locations, the Alfvén waves absorb energy from the alpha particles – helium nuclei that are produced by the fusion of hydrogen atoms. Alpha-particle energy is an essential component in keeping the fusion plasma hot enough to "burn."

At left, X-ray image of a long-lived Alfvén wave in the Wendelstein W7-AS stellarator, a magnetic-confinement device. The alternating red/yellow and purple regions represent crests and troughs of the wave. The gridded black lines depict a vertical cross-section of the plasma in the torus. Note that R measures the radial distance from the center of the torus, which is to the left; Z labels the vertical distance measured from a horizontal reference line that passes through the center of the torus.

Image courtesy of A. Weller.

Solar prominences are cool plasmas that float in a hot corona. The cloud-like prominence structure is located 10 000 to 20 000 km above the limb of the Sun (dark line in the middle of the frame). High-resolution images show wavy motion of such prominences. Solar physicists have attributed these observations from the satellite Hinode to Alfvén waves in the corona.

Image courtesy of T. J. Okamoto.

A Rough Ride from the Sun

WAVE ACTIVITY ALSO PLAYS AN IMPORTANT, if incompletely understood, role in the formation of turbulence in the solar wind – the titanic stream, comprising ions of hydrogen, helium, and a few other elements, that blasts outward from the Sun at speeds ranging (on average) from 200 to 600 kilometers per second.

The solar wind is one of the most difficult phenomena in plasma science to study because it is subject to a combination of chaotic and turbulent forces. For one thing, it originates in the solar corona, an area that is constantly fluctuating in density, electromagnetic properties, and energy content. Then, as the torrent of particles travels, it is subject to effects from Alfvén and other waves, including collisions between two or more waves.

In fact, a dominant form of turbulence in the solar wind is believed to originate when two Alfvén waves, proceeding in opposite directions while moving parallel to the magnetic field lines in a region, slam into each other. That causes a powerful release of energy in a direction basically perpendicular to the field lines.

Thanks to those and other effects, which can be detected by spacecraft, the solar wind's properties fluctuate on a stupefying range of time scales – from 27 days (the time it takes the Sun to make one revolution) to milliseconds. However, one of the few generalizations that *can* be made about its fluctuations is that they often seem to be Alfvénic in origin. One distinctive feature of Alfvén waves is that their velocity is directly correlated with the strength of the magnetic field, and this same pattern is evident in measurements of the solar wind.

Interestingly, because the solar wind is a fluid (made up of charged particles and therefore subject to electrical and magnetic fields, but a fluid nonetheless), it behaves in ways similar to the onset of turbulence in other fluid realms, such as liquids flowing in pipes or air that has traveled over an airplane wing. Progress in those areas may shed light on solar wind behavior, and vice versa.

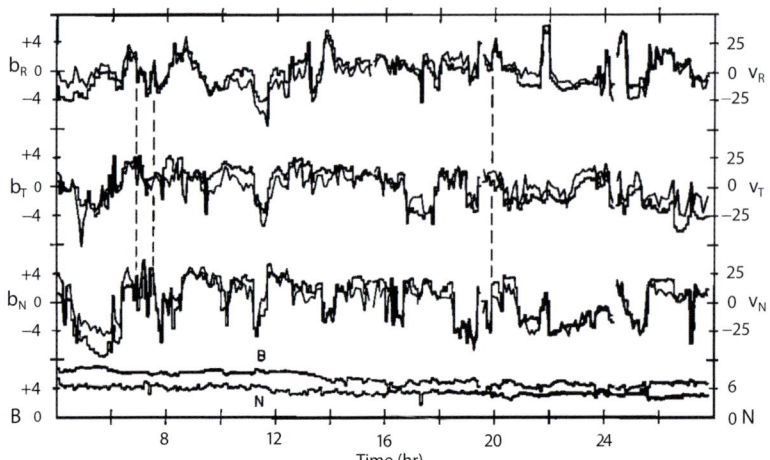

For an Alfvén wave, the three components of the velocity (v_R, v_T, v_N) and magnetic field (b_R, b_T, b_N) perturbations are strongly correlated. The two lower curves represent the background magnetic field strength B and the proton density N.

Image courtesy of J.W. Belcher and L. Davis. This image was published by AGU: *Journal of Geophysical Research* **376** (1971), no. 16; copyright 1971.

Hannes Alfvén
(1908–1995)

PLASMA PHYSICIST

NOBEL PRIZE, 1970,
for "fundamental work and discoveries in magnetohydrodynamics with fruitful applications in different parts of plasma physics."

Hannes Alfvén

HANNES ALFVÉN IS WIDELY CONSIDERED to be the father of magnetohydrodynamics – the study of the dynamics of electrically conducting fluids – and gave the field its name. He was born in Norrköping, Sweden to a family with high standards of achievement. Both his parents were physicians, and he numbered among his uncles a celebrated composer, a noted inventor, and a successful agronomist. As a boy, Alfvén developed two interests that would last a lifetime: electricity and electromagnetic waves (from an early enthusiasm for radio sets) and astronomy (from a book given to him as a gift).

By 1932, he had received his doctorate from the University of Uppsala and he began a career characterized by highly original ideas that were often initially rejected as unorthodox, only to be proven right later. His theories about the interaction of electrical and magnetic fields in plasmas were met with skepticism, but provided the fundamental insights underlying the discovery of the Van Allen radiation belts. And his conviction that electromagnetic waves could propagate in a conductive plasma, the mechanism at the heart of what are now called Alfvén waves, eventually provided an explanation for a host of terrestrial and astronomical phenomena. He first published a description of the waves in 1942, but his work was not widely accepted until six years later, when the famous physicist Enrico Fermi – after hearing a lecture on the subject by Alfvén at the University of Chicago – nodded and remarked "of course."

After a few years teaching at Uppsala and the Nobel Institute, at the extraordinarily young age of 32 Alfvén was named professor of electromagnetic theory and electric measurements at the Royal Institute of Technology. He later joined the faculty at the University of California at San Diego. Throughout his later career he elaborated theories about the nature of the galactic magnetic field and the role of plasmas in space, published several text books, and somehow found time to write a political and scientific satire titled "The Great Computer: A Vision."

Comet Hale-Bopp. The wake behind the comet is pointing away from the Sun and is caused by the solar wind.
Image copyright 1997 by Fred Espenak, www.MrEclipse.com.

The Sun–Earth Connection

THERE ARE 100 BILLION STARS in our galaxy, give or take a few billion, ranging from sputtering wimps barely larger than Jupiter to blazing monsters 2000 times the diameter of the Sun. What they have in common is that each one of them is an incandescent plasma, including our indispensable and boisterous local star.

The Sun shines because, deep inside it, a process called nuclear fusion combines nuclei of hydrogen, the most abundant element in the cosmos, to create nuclei of helium, the second most abundant. In doing so, it releases vast amounts of energy from tiny dimensions. This is possible because within the atomic nucleus (10 000 times smaller than the atom itself) resides the mightiest force in nature: the aptly named "strong" interaction that holds the components of nuclear particles together. One manifestation of that force, called "binding energy," keeps protons and neutrons – collectively known as nucleons – tied firmly to one another in the nucleus.

That's a stupendous feat. The elementary hydrogen nucleus is a single proton. Protons have a positive electric charge, and therefore naturally repel other protons. And that repulsive force increases as the particles get closer.

Still, it is no match for the binding energy, which is about 20 times stronger than the electromagnetic force. But the binding energy only takes over at *very* close range: it loses strength beyond a couple millionths of a billionth of a meter.

So in order to get nucleons to bind, they have to be squeezed together hard enough to overcome the electrical repulsion. The extremely high-energy collisions required are produced by very high temperatures, as high as 15 million K at the solar core. At those temperatures any material is in a plasma state, which is why fusion and plasma physics are so intimately linked. And, even then, the nucleons have to come together in precisely the right way, a situation that only occurs for only a fraction of time in even the staggering densities of stellar plasma.

Lighting up the Solar System

> BUT WHEN NUCLEONS FINALLY DO COMBINE, an amazing thing happens. The total mass of the resulting fused nucleus is just slightly *less* than the separate masses of its components. (It is the atomic analog of the fact that two people can live more cheaply as a couple than both could separately. In physics terms, one says that the fused nucleus has a lower potential energy than its constituent parts.) The difference is less than 1 percent, but it has a gigantic impact.

That's because the "missing mass" difference is given off as energy according to $E = mc^2$. Multiply even a minuscule quantity of mass (m) by the speed of light (c, about 300 000 kilometers per second) *squared*, and the product is a spectacularly large number. But fusion only gets things started. The Sun's appearance and often mysterious behavior involve several factors acting simultaneously.

Basically, most of what happens in and on the Sun is the result of radiation, convection, solar rotation, and electromagnetic interplay. Radiation of fusion-generated energy from the center of the Sun transfers heat outward. About two-thirds of the way from the solar core, the plasma density drops to the point at which the energy begins to power the same kind of convection patterns that form thunderstorms on Earth: hotter, less dense, matter rises until it cools, thickens, and descends back into higher-temperature regions where it heats up and rises again, creating a sort of conveyor belt. On Earth, convection in the atmosphere creates lightning, among other things. In the Sun, convection prompts hugely convoluted electromagnetic effects. Every particle in a plasma is electrically charged, and the motion of charges creates magnetic fields. At the same time, changing magnetic fields induce currents to flow, which in turn produce more magnetic fields, and so on.

The ensuing electromagnetic tangle would be complex enough by itself. But it is further complicated by the fact that the Sun rotates in a seemingly very odd way. It spins in the direction of the Earth's orbit at an *average* rate of about one revolution every 28 Earth days. However, its spin rate varies drastically depending on location, from 34 days at the poles to 27 days at the equator. Moreover, below the surface, layers of plasma at different depths revolve at different speeds.

Solar rotation and polar flows of the Sun as deduced from measurements from the satellite SOHO (Solar and Heliospheric Observatory). The cutaway reveals rotation speed inside the Sun. The left side of the image represents the differences in rotation speed between various areas near the surface of the Sun. Red-yellow is faster than average and blue is slower than average. The light orange bands are zones that are moving slightly faster than their surroundings. SOHO observations indicate that these extend down approximately 20 000 km into the Sun. Sunspots, caused by disturbances in the solar magnetic field, tend to form at the edge of these bands. Image courtesy of SOHO (ESA & NASA) Consortium.

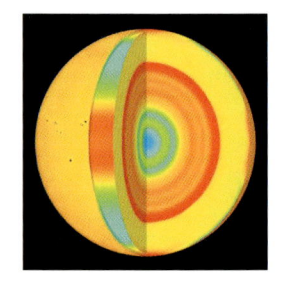

Spots, Loops, and Lariats of Fire

SOME DEEP SELF-GENERATING ACTION in this writhing jumble gives rise to the Sun's overall magnetic field, which reverses polarity every 11 years through an undetermined mechanism. Meanwhile, deep currents of moving plasma generate countless smaller, localized fields whose lines can protrude through the solar "surface," called the photosphere. (This surface is actually just the visible outer layer of plasma, a mere few hundred kilometers thick.) These fields originate thousands of kilometers below the photosphere, where they are stretched and twisted by the Sun's weird multispeed rotation. That displacement frequently causes a section of the field lines to bulge upward, often far above the photosphere, before plunging down again to complete the loop.

The results take many forms studied by plasma scientists. One is the familiar sunspot, a dark region in the photosphere that is typically 20 000 to 60 000 kilometers in diameter. As the celebrated American astronomer George Ellery Hale (1868–1938) determined, a spot represents the point at which magnetic field lines are poking either straight out or straight in through the photosphere. Each spot has two distinctive features: a dark central area, or umbra, where the field strength is highest (reaching a few thousand times the strength of Earth's planetary field); and a slightly brighter surrounding area, called the penumbra, where the field lines are less concentrated. Sunspots appear "dark" because at an average temperature around 3500 K, they are notably cooler than their radiant surroundings at 5800 K, and therefore less luminous. But even the darkest umbra is hotter than the filament of a household light bulb.

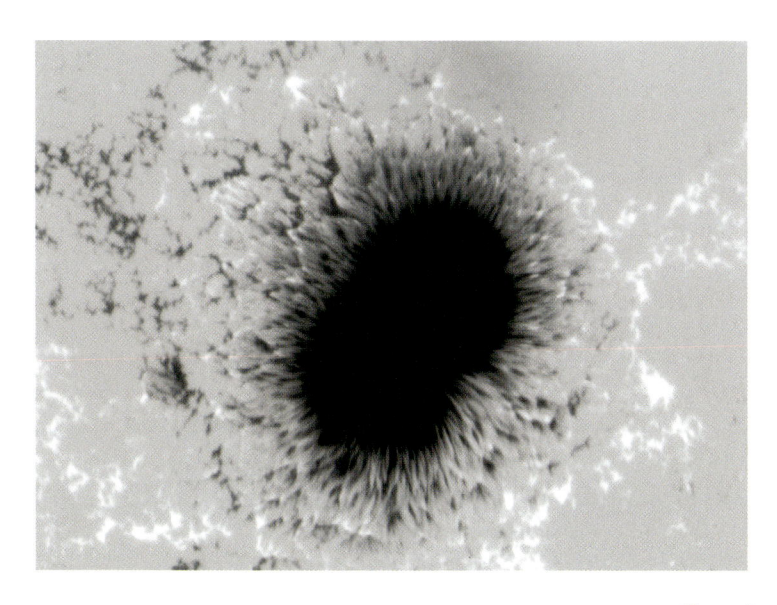

"Magnetogram" from the satellite Hinode showing the concentration of magnetic fields around a single sunspot in the solar photosphere.

Image copyright NAOJ / JAXA / NASA / SFC / ESA.

Hale showed that sunspots very often form in pairs of opposite polarity. One spot in the pair marks the area where field lines are bursting out of the photosphere, while the other shows where the lines descend again. Parts of the Sun with an abnormally large volume of pairs and other magnetic phenomena are called active regions, and are the source of many of the Sun's spectacular special effects. One is the solar loop, in which magnetic field lines connecting a sunspot pair arch upward, thousands of kilometers above the photosphere, bearing streamers of entrained plasma that light up the neighborhood like ropes of flame.

Some particularly vigorous loops can extend all the way through the chromosphere (the next solar layer out from the photosphere) and into the next and last layer – the tenuous corona, an ultra-rarefied plasma that stretches millions of kilometers into space. By comparison to the photosphere, the corona's visible light is so faint that it can only be seen during a total solar eclipse. Common sense suggests that the coronal glow is dim because it is so far from the Sun's central energy source, and therefore must be very cool. On the contrary, the corona is incomprehensibly hot, on the order of 1 to 3 million K. That's 100 to 300 times hotter than the thin chromosphere beneath it, and 200 to 600 times hotter than the photosphere. One of the most stubborn problems in plasma science involves explaining the mechanisms that produce such exceptional anomalous heating.

One leading candidate is Alfvén waves (discussed in Chapter 2). Another possibility involves the action of "nanoflares." It has been known since the 1970s that small flashes of energy, lasting only a few seconds, are common all over the lower corona. Like their larger, longer-lived cousins – the notorious solar flares – nanoflares are presumed to arise from energy released when two magnetic tubes intersect, produce intense concentrations of electrical current density, and "reconnect" in a sort of splicing action that liberates a burst of energy. If enough of that energy is transferred to electrons and ions in the coronal plasma, it might account for a substantial part of the heat.

On the facing page, the graph in the left column shows the variation in the plasma temperature of the Sun as a function of altitude (in kilometers) in the upper atmosphere above a typical sunspot. Note the rapid increase in the temperature over a narrow vertical region between 1000 and 10 000 km. The green band shows the image from Hinode of a sunspot in the photosphere (based on radiation of wavelength 430 nanometers). The blue band is an image of the same sunspot viewed from the chromosphere (based on radiation of wavelength 396 nanometers). The yellow band represents an image of the transition region, where the temperature rises rapidly, viewed from TRACE. The orange band represents an X-ray image of the corona at higher altitude, viewed from Hinode.

Image copyright NAOJ / JAXA / NASA / SFC / ESA.

NEW VIEWS OF THE SUN FROM HINODE AND TRACE

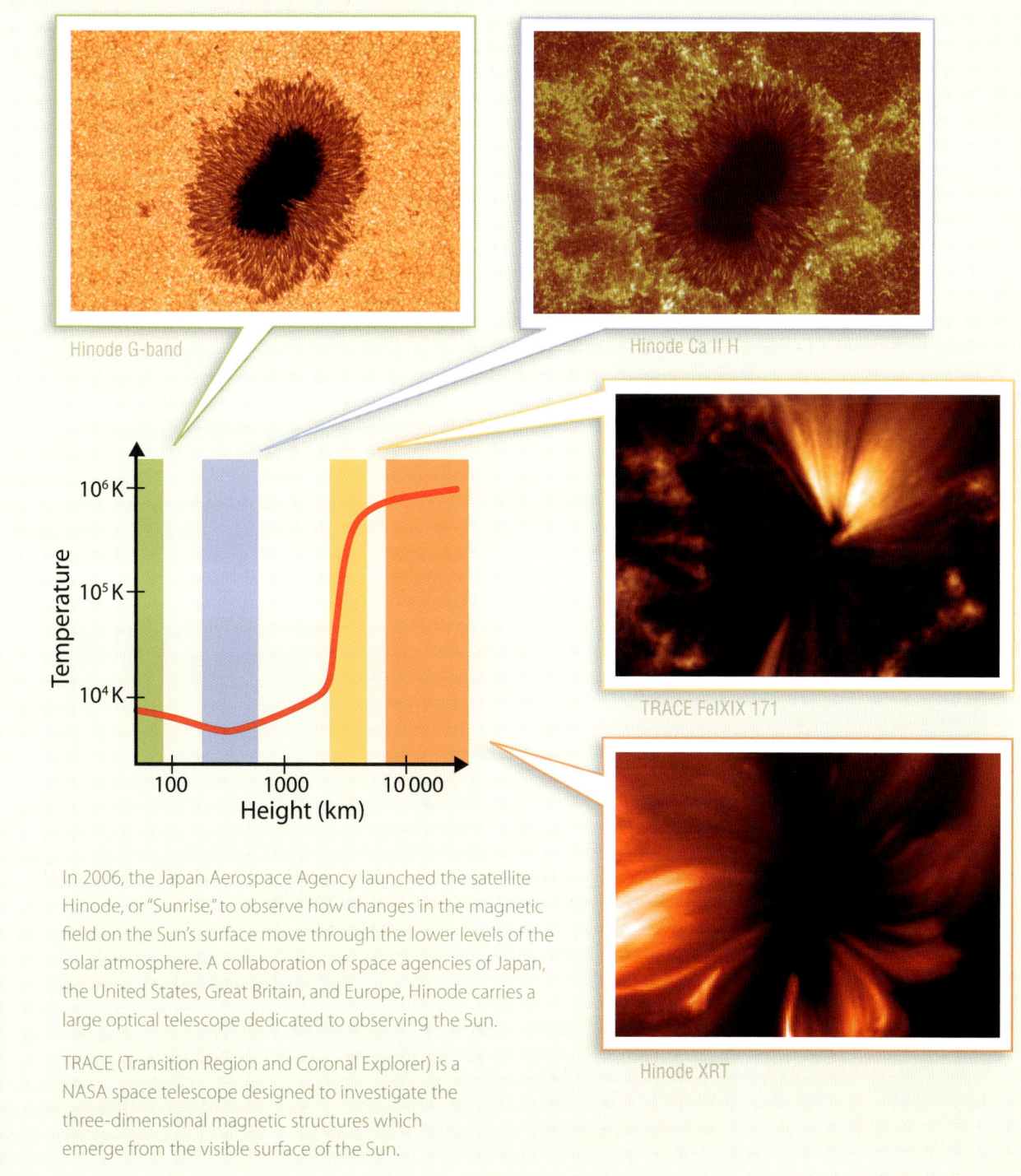

Hinode G-band

Hinode Ca II H

TRACE FeIXIX 171

Hinode XRT

In 2006, the Japan Aerospace Agency launched the satellite Hinode, or "Sunrise," to observe how changes in the magnetic field on the Sun's surface move through the lower levels of the solar atmosphere. A collaboration of space agencies of Japan, the United States, Great Britain, and Europe, Hinode carries a large optical telescope dedicated to observing the Sun.

TRACE (Transition Region and Coronal Explorer) is a NASA space telescope designed to investigate the three-dimensional magnetic structures which emerge from the visible surface of the Sun.

THE SUN–EARTH CONNECTION

Blast from the Mass

THE MAGNETIC FIELD in the Sun's corona represents an enormous reservoir of stored energy. To understand just how much energy is available, the violent electromagnetic eruptions called solar flares provide a gauge of astronomical proportions. Flares are bursts of radiation that occur suddenly, with no prior indication, as often as dozens of times a day at the peak of the sunspot cycle. In a matter of seconds, a volume about the size of the Earth (one-hundredth of the solar diameter) reaches tens of billions of degrees, emitting gamma-ray, X-ray, and ultraviolet radiation briefly, totaling nearly one-tenth of the Sun's total output – equivalent to tens of billions of the atomic bombs dropped on Hiroshima.

Why? It certainly isn't heat transfer. There isn't enough thermal energy in the entire corona to account for that sort of outburst. So what invisible process could possibly amass that kind of power, and release it so swiftly? Plasma physicists have struggled for decades to answer that question, and many believe that magnetic "reconnection" is a solution.

Flares typically appear around and above active regions and sunspot areas where magnetic fields are especially strong. Magnetic field lines form closed loops; like everything else in the cosmos, those loops naturally tend to occupy the lowest-energy condition – in this case, the shortest length and most symmetrical shape possible. But the incessant churning and differential rotation of plasma layers in the Sun places powerful torques and shear forces on the field lines, stretching and deforming them into contorted curves. The more energy applied to bend them, the more energy they contain. In the same way, the harder an archer pulls back on the bow string, the more energy is stored for the shot.

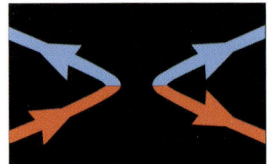

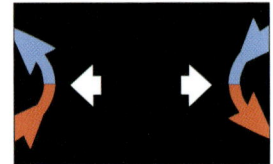

At right, a false-color image of coronal loops taken with NASA's Transition Region and Coronal Explorer satellite. Image courtesy of NASA.

The figure below illustrates the magnetic reconnection process when two magnetic fields pointing in opposite directions merge. The field lines break, reconnect, and pull away after the reconnection process is complete. Image reprinted by permission from Macmillan Publishers Ltd: P. Hanlon et al., *Nature Physics* (Aug. 1st, 2005), doi:10.1038/nphys111; copyright 2005.

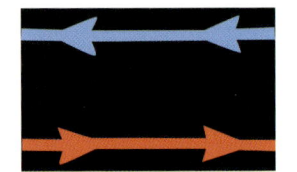

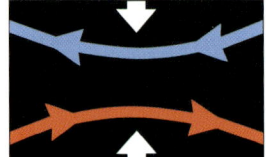

So when two or more of these highly stressed loops (with opposite field directions) come into contact – typically in the lower corona – they merge and re-form into lower-energy configurations in the process called reconnection, releasing the now-surplus energy much as the bow string transfers its energy to the arrow. The release propels electrons and protons outward at very high speeds, generating the high-frequency photons of the flare.

Interestingly, it turns out that "breaking" and "reconnecting" of field lines cannot occur if the plasma is a perfect conductor. However, if the plasma has even a small amount of resistance, then the loops can touch in a small volume, called the "diffusion region," where the field lines can reconnect. This process liberates energy, causing jets of plasma to fly away from the reconnection point and into the surrounding area. In flares, however, the energy is released much faster than current theory can account for, and the process remains the subject of intense study.

Whatever drives the flares, it is ferociously powerful: solar flares occur 150 million kilometers from Earth, but their effects can be felt strongly enough here to turn long-distance radio and satellite communications into a snarl of static and to force astronauts in the International Space Station to hunker down until the radiation subsides. In 2003, an uncommonly strong flare dumped so much energy into the Earth's atmosphere that the Northern Lights were seen as far south as the Mediterranean.

Even more worrisome to our home planet, however, are the epic ballistic events called Coronal Mass Ejections (CMEs). Near the peak of the 11-year solar cycle, CMEs occur about once or twice a day, apparently powered by magnetic forces and expending approximately the same amount of energy as a solar flare. But in the case of CMEs, nearly all the energy goes into propelling tens of billions of tons of plasma outward at hundreds of kilometers per second, tearing a huge hole in the corona. If the ejection is pointed in our direction, the expanding mass reaches Earth in about four days, carrying its magnetic structure with it.

First observed in the early 1970s by solar-detection spacecraft, CMEs are now closely monitored by numerous instruments that track "space weather." Advance warning can be critical: in 1989, induced current spikes from a giant CME blew out HydroQuebec's electrical grid, leaving six million people without power for half a day.

Image of an eruption from the Sun when a large clump of solar mass is ejected into space, viewed from the satellite SOHO. On the scale of the image to the right, the Earth is the size of this blue sphere ●.
Image courtesy of SOHO.

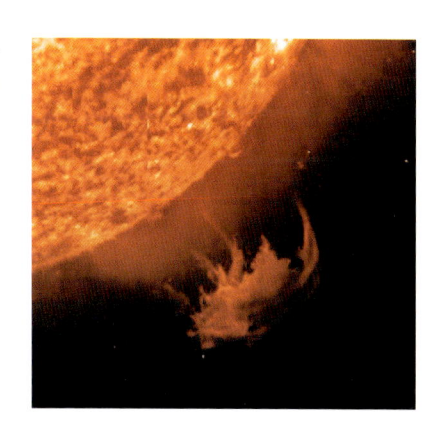

A Mighty Wind

FLARES AND CORONAL MASS EJECTIONS, HOWEVER, do not begin to exhaust the roster of powerful coronal phenomena. Another is the production of the "solar wind," an immense, continuous stream of electrons and protons that blows off the corona in all directions at the rate of about a billion kilograms per second and extends all the way to the end of the solar system and beyond. It occurs because, although the average temperature of particles in the corona is between 1 million and 3 million K, that's *only* an average: some tiny fraction of the coronal miasma is always hot enough – that is, moving fast enough – to escape the Sun's gravitational pull. Those escapee particles form the solar wind, the largest plasma structure in the solar system. It has been blowing for about 5 billion years, but it is only in the past few decades that scientists have come to understand it.

The first evidence for this ghostly gale was seen centuries ago, when observers noticed that the tails of comets always pointed away from the Sun *(see page 26)*. By the mid 1950s, scientists had speculated about a solar "corpuscular radiation" that put pressure on the comet tails, and had calculated that the corona must remain very hot at extraordinary distances from the Sun. In 1958, Eugene Parker devised a new, comprehensive theory of what he called the "solar wind" – a stream of high-energy particles that expands from the corona, gaining speed as it moves farther from

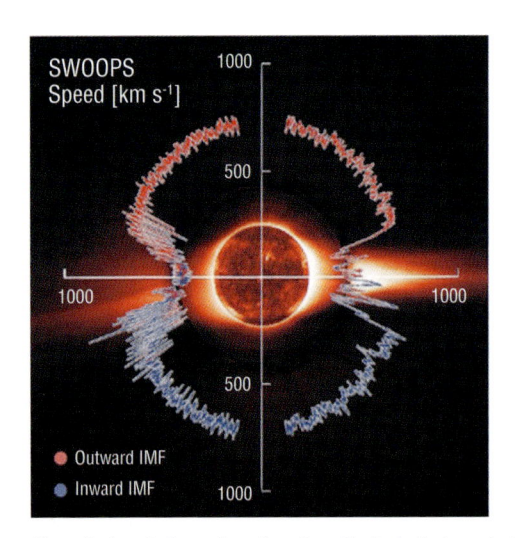

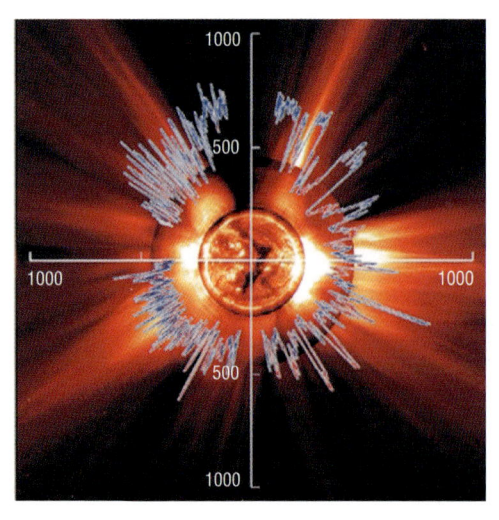

Plots of solar wind speed as a function of latitude during solar "minimum" when the Sun is relatively quiet (left image) and solar "maximum" when the Sun is very active (right image). Note that the solar wind speeds tend to be slow near the equatorial plane and fast in the polar regions. The acronym IMF denotes the interplanetary magnetic field. SWOOPS is a detector on board the sapcecraft Ulysses.

Image reprinted by permission from D. J. McComas. This image was published by AGU: D. J. McComas *et al.*, *Geophysical Research Letters* **30** (2003), no. 10, p. 1517, doi:10.1029/2003GLO017136; copyright 2003.

the Sun's center of mass, traveling across the solar system at about 400 kilometers every second. Within a few years, Soviet and American space probes returned evidence of such a supersonic plasma torrent, confirming the theory and establishing solidly the concept of the solar wind.

Fifty years later, the solar wind is the subject of continuing research. For one thing, its trajectory reflects the electromagnetic environment in the solar system, including the field lines that demark the interplanetary magnetic field. For another, its interaction with our planet's magnetic field has important – and often vividly visible – results.

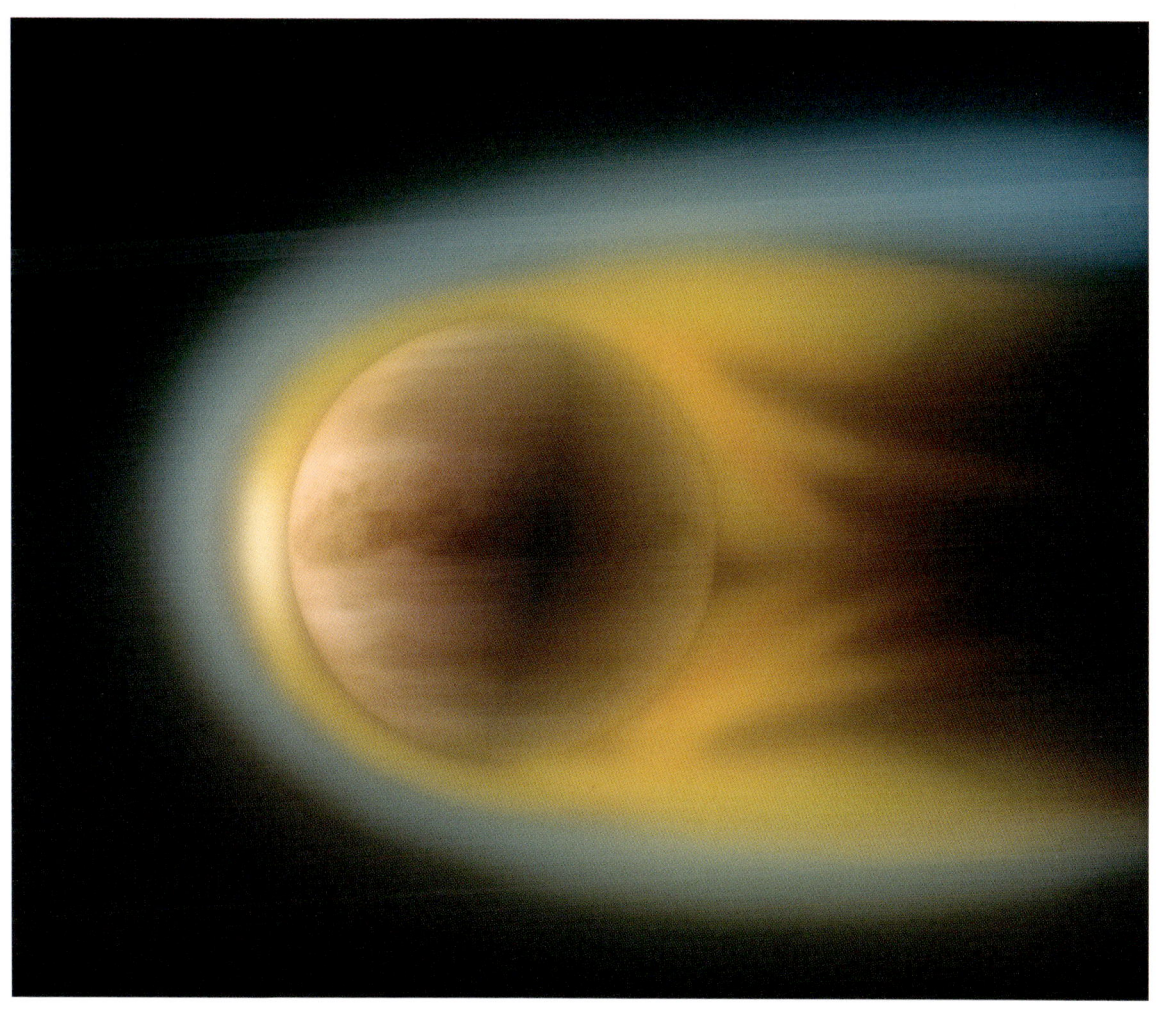

Interaction of the Venus atmosphere with the solar wind. Compared with the Earth, Venus has a much weaker (possibly even zero) magnetic field, so that the solar wind interacts directly with the atmosphere, much like it interacts with a comet. Image courtesy of European Space Agency.

BLOWING IN THE WIND

Earth, like Mercury and the gas giants, has a magnetic field of its own that extends about 126 000 kilometers into space. So when the million-mile-per-hour solar-wind plasma slams into the terrestrial field, called the magnetosphere, it is impeded and diverted such that it flows around the planet, like water around a rock in a stream. (This situation protects the surface of the Earth from electron and ion bombardment, with salutary consequences for the evolution and protection of life in these parts.) Over the years, plasma scientists have determined that by the time the solar wind arrives here, its density is about 10 million particles per cubic meter. That may sound like a lot, but it is a billionth of a billionth of the density of air at sea level. And only about one out of 1000 particles in the solar wind makes its way inside the Earth's magnetic field.

Of course, even that tiny fraction produces some impressive effects. For example, high-energy electrons and protons are trapped in two concentric, donut-shaped regions, called the Van Allen radiation belts in honor of their discoverer, which surround the Earth's equator. One notably low-lying region in the inner belt, known as the South Atlantic Anomaly, is particularly dangerous: satellites that pass through it have to be radiation-hardened, and the Hubble Space Telescope's instruments are deactivated during transit there. The solar wind also powers aurorae, the glittering curtains of light that form at high northern and southern latitudes when high-energy electrons from the solar wind collide with the upper atmosphere. These electrons transfer energy to oxygen and nitrogen molecules, which release the energy as photons in their famous ethereal glow *(see page 3)*.

A diagram of the Earth's magnetosphere – the region controlled by the Earth's magnetic field and shaped by interaction with the solar wind. The wind compresses the Earth's magnetic field on the day side (the side of Earth facing the Sun) to form the bow shock, and drags out the magnetic field on the night side (the side of Earth away from the Sun) to form the magnetotail, like the wake of a ship. Thus, the magnetosphere effectively shields us from energetic particles that are present in the solar wind or originate from cosmic sources. Within the Earth's magnetosphere, a magnetic cavity holds the Van Allen radiation belts. These are belts of energetic charged particles trapped by the Earth's magnetic field.

Image courtesy of NASA /Marshall Space Flight Center.

Thanks to advances in plasma physics, scientists now have a relatively complete idea of how solar-wind particles manage to get inside the magnetosphere – and it can seem highly counterintuitive. Usually, most of the penetrating particles do *not* enter at the initial collision, or "bow shock," where the wind first smashes into the Earth's field. Instead, it happens on the far lee (night) side of the planet, out where the plasma flow tapers down to a narrow wedge in an area called the magnetotail. As the magnetic field lines are drawn into proximity there, they can merge in the same process of reconnection that presumably releases energy in solar flares. In this case, the particles are driven in two opposite directions: outward away from the Earth; and directly back toward the planet via a pathway called the plasma sheet. This process is greatly enhanced if the field polarity of the solar wind is opposite to the Earth's magnetosphere, which makes it easier for field lines from each system to converge.

COLLISION COURSE

Naturally, all solar–magnetosphere interactions are magnified during the highest activity point in the 11-year solar cycle, when flares, CMEs, and solar-wind disturbances are most frequent – and terrestrial consequences most dire. The strongest of these events, called geomagnetic storms, occur scores of times per year at solar maximum, persist for one to several days, and cause worldwide effects.

For example, a large CME can squash the day-side magnetosphere to about half its normal depth, deranging navigational systems and threatening satellites. Or smaller substorms can form, typically near the poles, intensifying the aurorae for hours at an altitude of around 100–400 kilometers. And any process that adds energy to the atmosphere will cause the air to expand, thus increasing drag on orbiting objects and decreasing their altitude.

Analogous phenomena occur around other planets, such as Jupiter, which has a field strength ten times higher than Earth's and a magnetosphere as large as the Sun. But to realize the extreme limits of space plasma behavior, it is necessary to look far outside our solar system, to objects on size and energy scales that stagger human imagination.

Composite image of Jupiter from the Chandra satellite superimposed on an image from the Hubble Space Telescope. Note the X-ray activity near the poles where Chandra is detecting auroral lights due to the interaction of the solar wind with sulfur and oxygen ions trapped in Jupiter's magnetic field. These aurorae are 1000 times more intense than those seen on Earth.

X-ray image courtesy of NASA/CXC/SwRI/R.Gladstone *et al.*;
optical image courtesy of NASA/ESA/Hubble Heritage (AURA/STScI).

Eugene N. Parker
(born 1927)

PLASMA ASTROPHYSICIST

NATIONAL MEDAL OF SCIENCE, 1989
MAXWELL PRIZE IN PLASMA PHYSICS, 2003

Eugene N. Parker

IN TODAY'S WORLD, where hundreds of satellites orbit the Earth and elementary-school students look at pictures of our planet taken from the moon, it is difficult to imagine how Eugene Parker could have made so many fundamental contributions to plasma astrophysics before the advent of the space age. Born in Houghton, Michigan, Parker received his Ph.D. from Caltech in 1951 – six years before the Soviet Union launched a small satellite called Sputnik, and changed history, as well as Parker's future.

Parker spent four years at the University of Utah, and moved to the University of Chicago in 1955, where he and colleagues would gain notoriety by questioning the widespread dogma that space between the planets was basically an evacuated void populated only by the occasional stray particle or cosmic ray. When he began work on the idea, there were very few relevant observations. Although there were two principal existing theories of particle emanation from the Sun, neither was convincing.

In 1958, Parker, employing principles of plasma physics, limited data, and strong conviction, announced a theory of a surprisingly dense solar wind. Because the corona has such a high thermal conductivity, he reasoned, temperatures would remain high out to a point at which the Sun's gravitational force would be too weak to hold the hottest plasma particles. They would escape in a supersonic flow, and would take on a distinctive spiral shape (sometimes compared to the spray from a lawn sprinkler) as the wind propagated outward through the solar system.

Reaction was immediate and negative. Two reviewers rejected an article he submitted to the prestigious *Astrophysical Journal*, though the editor, Nobel Laureate S. Chandrasekhar, overruled them. But confirmation was on the way. Within months, in the post-Sputnik frenzy of satellite launches, the Soviet Union's Luna satellites detected evidence of such a solar wind. These findings were placed on a definitive foundation a few years later by the US Mariner 2 spacecraft.

Parker subsequently received the National Medal of Science and was elected to the National Academy of Sciences. His influence continues today, both through important books such as *Cosmical Magnetic Fields* and through the example he provides for innovative young scientists.

Bringing the Sun to Earth:
The Story of Controlled Thermonuclear Fusion

SINCE THE DAWN OF CIVILIZATION, mankind has looked to the Sun as the ultimate sustainer of life. And now plasma physics is on the verge of capturing the power of our star – nuclear fusion – and bringing it to Earth in a controllable form.

That effort, one of the most arduous in the history of science, requires that human beings actually beat the Sun at its own game. Researchers have to *outshine* the Sun by using temperatures that are ten times hotter. They have to employ isotopic fuels that are rare but much more efficient than the bare protons that fuse in the solar core. And they have to contort the infernal plasma into specific shapes and exacting densities while confining it in a safe enclosure. Those goals are so formidably hard that half a century of research has not yet produced a working reactor. Still, the long struggle continues, because the potential payoff could change the world.

The reason is that nuclear reactions – fission and fusion – are sources of immense power. And pound for pound, fusion produces many times more energy than even nuclear fission, the source of power for "atom bombs" and today's nuclear electrical plants.

For example, the energy released in fission when uranium-235 splits into smaller nuclei is about 0.8 million electron volts (MeV) per nucleon.

By comparison, consider the yield from a fusion reaction when one ion of deuterium (a heavy isotope of hydrogen, containing one proton and one neutron in its nucleus) and one ion of tritium (another hydrogen isotope with two neutrons) fuse into helium and liberate a neutron. The total energy, carried by the helium nucleus and the neutron, sums to 17.6 MeV. That's over 3.5 MeV per nucleon, nearly five times higher than the fission yield, which was already huge.

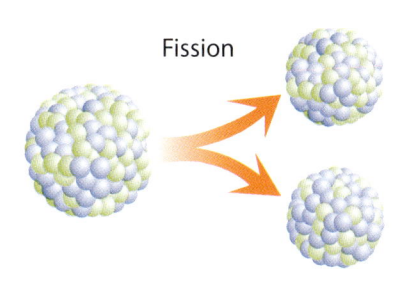

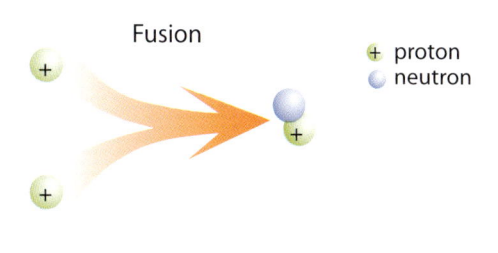

The magnitude of fusion reactions was conclusively revealed in test-firings of "hydrogen bombs" in the middle of the twentieth century. So were its stunning peacetime possibilities. Fusion was immediately recognized as an enormously attractive, perhaps ideal, power source for electrical power generation. And a comparatively clean one: fusion does not produce long-lived, radioactive breakdown products like fission, creates no greenhouse gases like fossil-fuel combustion, and uses as its principal raw materials deuterium, which is easily extracted from water in virtually unlimited quantities, and tritium, which can be produced from lithium found in the Earth's crust. Physicists got to work.

After decades of effort, however, taming fusion has proven to be a great challenge. A sustained reaction depends on the simultaneous combination of three requirements so demanding that some stars barely get away with it: high temperature at the plasma core; sufficient density of particles at desired locations; and suitably extended periods of confinement in the optimal plasma configuration.

Temperature is the least troublesome criterion, despite the shocking numbers involved. The Sun is able to generate fusion at a mere 15 million K because the atoms in its plasma are subject to the crushing compression of gravity. (The solar interior, where about 100 billion billion billion atoms fuse every second, is so dense that a photon produced at the center of the Sun takes as much as a million years to ricochet its way to the surface and emerge as sunlight.) On Earth, where that sort of compaction is impossible, temperatures around 100 million K are necessary. Remarkably, researchers have reached, and even exceeded, that level repeatedly. Eventually, functional reactor plasmas may be two or three times as hot.

Controlled reaction and prolonged confinement, however, are much tougher targets, and researchers are pursuing two overall strategies for achieving them: magnetic and inertial.

Deuterium nuclei, consisting of one neutron (blue) and one proton (green), fuse with tritium nuclei, consisting of two neutrons and one proton, to produce a helium nucleus, consisting of two protons and two neutrons, and a neutron. The helium nucleus and the neutron are liberated with energies of 3.5 MeV and 14 MeV, respectively.

Image revised from Physics 2010 Committee, *Plasma Science: Advancing Knowledge in the National Interest* (Washington, D. C.: The National Academies Press, 2007), p. 19; with permission from ITER.

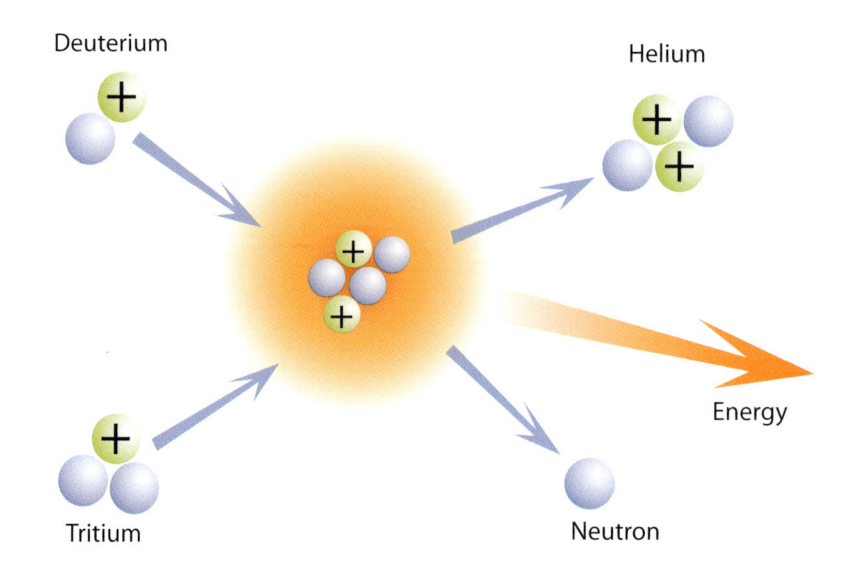

Magnetic Bottles

BECAUSE PLASMAS ARE MADE UP OF CHARGED PARTICLES, they can be manipulated by electrical or magnetic fields or both. So naturally one of the earliest concepts for snaring a fusion plasma so that it could "burn" in a limited volume without touching – and thereby destroying – its enclosure was to imprison it in a cage of powerful magnetic fields. Laws of physics dictate that the ions and electrons are constrained to travel along magnetic field lines; and they spiral around the lines as they go. The stronger the field, the smaller the distance that particles can stray from the field lines. Reactors have main fields that reach 10 or more tesla (T), a unit of high field strength. (By comparison, the Earth's magnetic field is about 0.00005 T.) The prodigious grip of those fields makes it possible to heat an experimental plasma to the desired temperature without it expanding enough to melt its enclosure.

Many designs have been tested. But the most dependably promising to date is a donut shape, or torus, which is achieved in two stages and shown in the figure below. Super-strong electromagnets, indicated as blue coils, generate a sheath of circular magnetic field lines like a horizontal stack of hula-hoops. A typical toroidal magnetic field line points in the direction of the blue arrow. Then an additional set of magnets, indicated as green coils in the "donut hole," are applied to induce an electric current in the highly conducting plasma. This electric current, which points in the direction of the green arrow, produces "poloidal" magnetic fields at right angles to the first set of toroidal magnetic fields. The toroidal and poloidal magnetic fields combine to produce a helical magnetic field which confines the plasma, indicated in semitransparent yellow, in the donut.

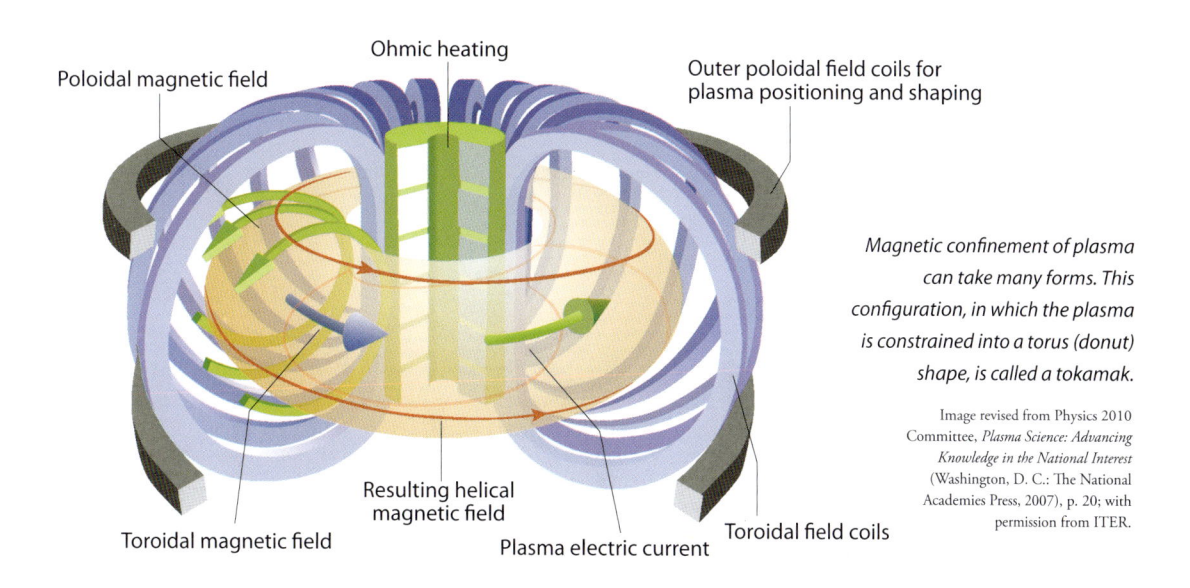

Magnetic confinement of plasma can take many forms. This configuration, in which the plasma is constrained into a torus (donut) shape, is called a tokamak.

Image revised from Physics 2010 Committee, *Plasma Science: Advancing Knowledge in the National Interest* (Washington, D. C.: The National Academies Press, 2007), p. 20; with permission from ITER.

Reactors with this general arrangement are called *tokamaks*, from an acronym of Russian words meaning "toroidal chamber with magnetic coils." More than a dozen are now in operation worldwide; and the most advanced – a multi-billion-dollar behemoth called the International Thermonuclear Experimental Reactor (ITER) – is under construction in France, with completion scheduled for some time around 2016. ITER is expected to reach 500 megawatts of total fusion power, comparable in output to today's medium-size fission reactors or hydroelectric and petroleum-based power plants.

Ingenious as the tokamak design is, it has an inherent liability. Ideally, the device should run continuously in a "steady state," with plasma current constantly circling the torus in the same direction to generate the confining poloidal magnetic field. This runs afoul of a familiar principle from high-school science: just any old magnetic field will not induce electrical currents; they are prompted to flow only when conductors are exposed to *changing* magnetic fields. (This happens 24/7 outside the average home as transformers use changing current flows in a primary coil to induce currents in a secondary coil. That alternating current, of necessity, reverses its direction cyclically – about 60 times a second in household circuits.) One way to keep changing the magnetic field is to run in "pulsed" mode, during which the field is generated, increased, decreased, and then re-increased in the same orientation. But this is inefficient.

Direct, not alternating or pulsed, current is essential for steady-state plasma confinement. So plasma scientists came up with a cunning "current drive" method: the plasma is exposed to electromagnetic waves of a few gigahertz (GHz) that are carefully adjusted so that the waves' electrical fields are parallel to the current flow. Electrons encountering those fields get a shove in the direction of the current just as surfers get a velocity boost when catching a wave.

Cutaway drawing of the International Thermonuclear Experimental Reactor (ITER), which is a tokamak. This multinational collaboration is expected to be commissioned in 2016. The small blue human figure at bottom center indicates the scale.

Image courtesy of ITER.

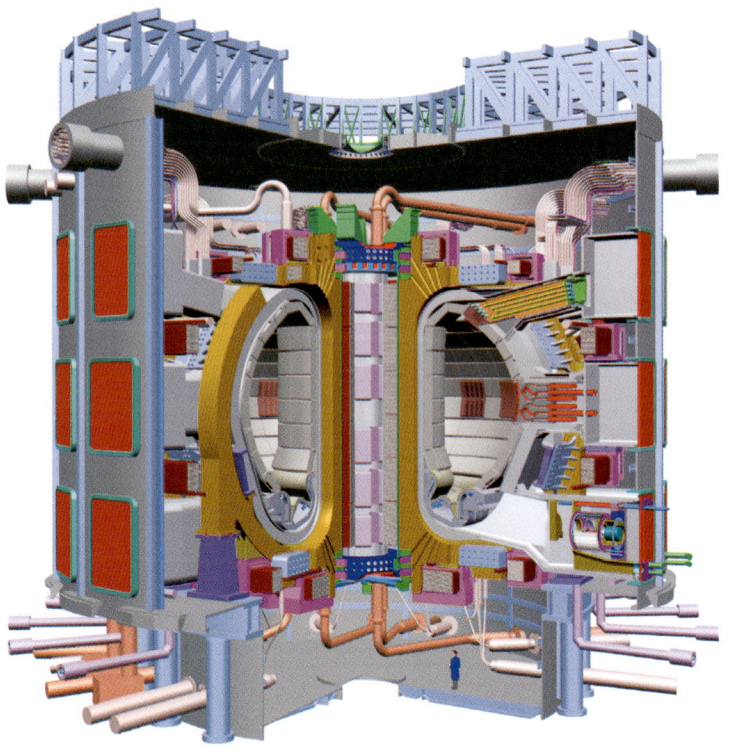

FIRING UP

There is no single technique that can raise the temperature of a tokamak to the fusion level, so multiple methods are used. First, the powerful electrical current induced by the magnets heats the plasma through which it flows by the same process of ordinary electrical resistance that makes a toaster work. This approach, called ohmic heating, has a built-in, self-limiting problem: as noted in Chapter 1, the electrical resistance of a plasma decreases as it gets hotter. Consequently, ohmic heating tends to peak out around 30 million K, and a second or third additional method is typically required.

One is to blast the plasma with neutral atoms traveling at high speeds. As they are trapped and ionized, these atoms transfer their kinetic (motion) energy to the plasma particles through collisions. In addition, scientists zap the plasma with radio-frequency waves of 100 to 300 GHz generated by a source outside the chamber. That radiation is tuned to precisely the frequency at which a large fraction of the plasma particles orbit around magnetic field lines, analogous to the way a kitchen microwave oven is optimized for rotating water molecules in food. The applied radiation can be very precisely localized: the frequency required to make the particles resonate is directly dependent on the strength of the surrounding magnetic field. And since the field strength varies substantially – but predictably – between the plasma core and the outer edges, researchers can generate just the right frequency for the section they wish to energize.

In addition, the plasma is heated by the energy of the reaction products themselves. At the instant of fusion, hydrogen isotope ions combine into a helium nucleus plus a fast-moving neutron. The helium nuclei carry off about 20 percent (3.5 MeV) of the energy released, further heating the plasma when they smash into nearby ions.

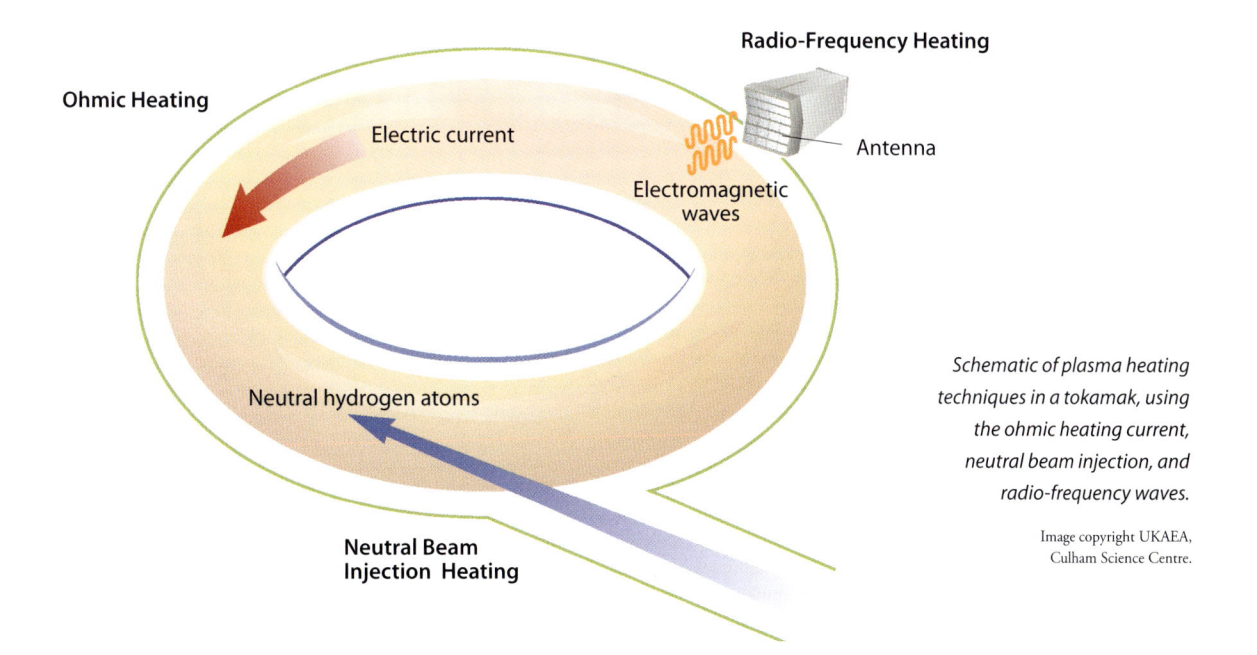

Schematic of plasma heating techniques in a tokamak, using the ohmic heating current, neutral beam injection, and radio-frequency waves.

Image copyright UKAEA, Culham Science Centre.

Meanwhile, the uncharged neutrons, bearing the other 80 percent (14.1 MeV), are unaffected by the magnetic fields. They fly away to strike a wall outside the plasma, imparting their energy to the wall material as heat. That heat, in turn, can be used to produce steam and turn turbines for electrical power generation in the same way conventional nuclear or fossil-fuel plants do.

To date, however, scientists have not been able to create a fusion reaction that produces more than about two-thirds of the power that it takes to start it, or that persists for more than a few seconds. The reasons can be found in the complex and obstinate peculiarities of plasma behavior.

UNRULY REACTIONS

In a fusion plasma, the hottest and densest concentration of particles tends to occur at the center of the torus, bound by the imposed magnetic field lines, with a smooth gradation of declining temperature and pressure toward the outermost edge. Operators could "tune" reactor performance on the fly by varying the amount of neutral atoms injected, the size of the induced current, and a few other adjustments.

But no matter how careful the control, hot particles move in non-uniform ways – and moving charged particles create their own magnetic fields, which in turn interact with the surrounding fields, distorting temperature, density, and pressure patterns, and re-routing field lines. Differences in current flow between neighboring sections can "tear" the plasma structure. Energy can bleed away from the core through radiation and/or instabilities. The plasma can react in various ways with the container wall. The myriad forms of resulting mayhem are complex. By means of sophisticated mathematical models and simulations on top-tier supercomputers, physicists are gradually learning how to avoid or correct many of the most worrisome conditions.

For example, one class of defects involves deviations from the initial orderly pattern of magnetic field lines and layers. "Magnetic islands" a few inches in width can form in the plasma, causing detours in field lines and

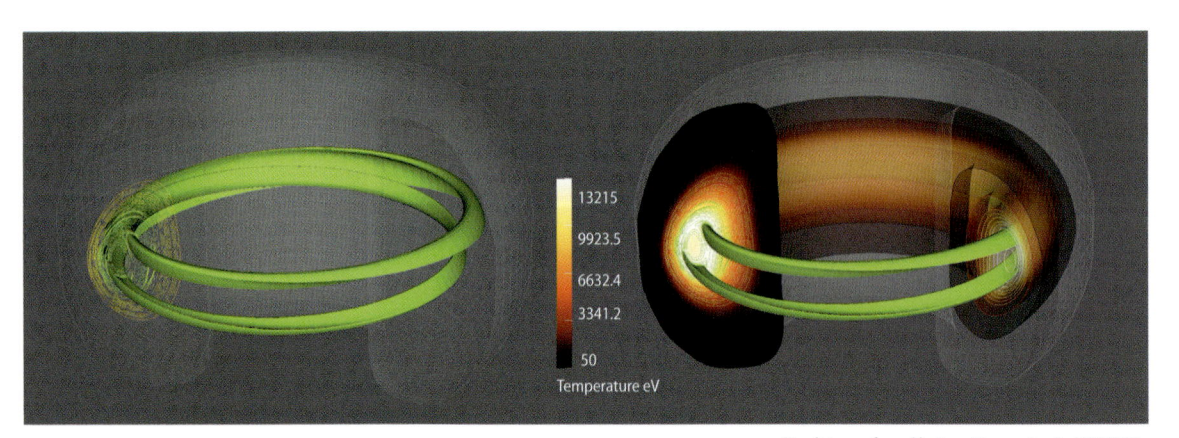

Computer simulations of the DIII-D tokamak reproduce many of the observed features, including energy confinement times and energy loss patterns in the presence of magnetic islands. The time scale of the energy loss is controlled by multiple islands that form spontaneously as the plasma discharge evolves. Shown here are the temperature contours that begin to change as the magnetic islands, shown on the left, form.

Simulation performed by Scott Kruger using the NIMROD code on flagship computers at the National Energy Research Scientific Computing Center (NERSC). Visualization and analysis performed by Allen Sanderson using the SCIRun Problem Solving Environment. Support provided by the US Dept of Energy Scientific Discovery Through Advanced Computing (SciDAC) Project.

their associated current flow. The resulting contact between formerly separate adjacent regions produces a form of the magnetic reconnection process described in Chapter 3. The shifted field lines can link hot areas with cooler ones, leading to heat loss in core areas and derangement of field structure. Although such islands can arise in a fraction of a second, researchers are learning to shrink them in real time by applying microwaves. The waves drive currents in the islands that replace those the islands lose when field lines shift, reducing island width by more than 50 percent in less than a second.

Another kind of headache results from turbulence, the same phenomenon seen in a stream when smoothly flowing water hits one side of a protruding rock and causes whirling eddies and gurgling bubbles on the downstream side. A plasma can act in much the same way. Unfortunately, the onset and evolution of turbulence is one of the most notoriously intractable problems in the history of science. Amazingly, plasma physicists have begun to understand and control it in tokamaks.

A usable sustained fusion reaction will have to rely on minimal transport of heat from the central plasma to the cool periphery. The present lack of such "insulation" is a horrendous obstacle, though its chief cause is increasingly clear: centimeter-scale pockets of microturbulence begin to form when the temperature difference between two adjacent plasma regions is above a certain threshold. (In much the same way, eddies form in a current between two volumes of water moving at different speeds.) Those pockets facilitate a devastating amount of heat exchange, as new imaging techniques have confirmed.

But recent research points to ways of stifling those effects. Scientists have created mathematical descriptions for the onset of plasma turbulence – an essential prerequisite to control – and have found that specific flow patterns can minimize turbulence. And experiments have revealed that, under certain conditions, plasmas spontaneously form boundaries called "transport barriers" that have drastically reduced turbulence and correspondingly good insulation properties. Knowing what conditions produce the barriers will allow physicists to adjust the properties of their plasmas to encourage barrier formation.

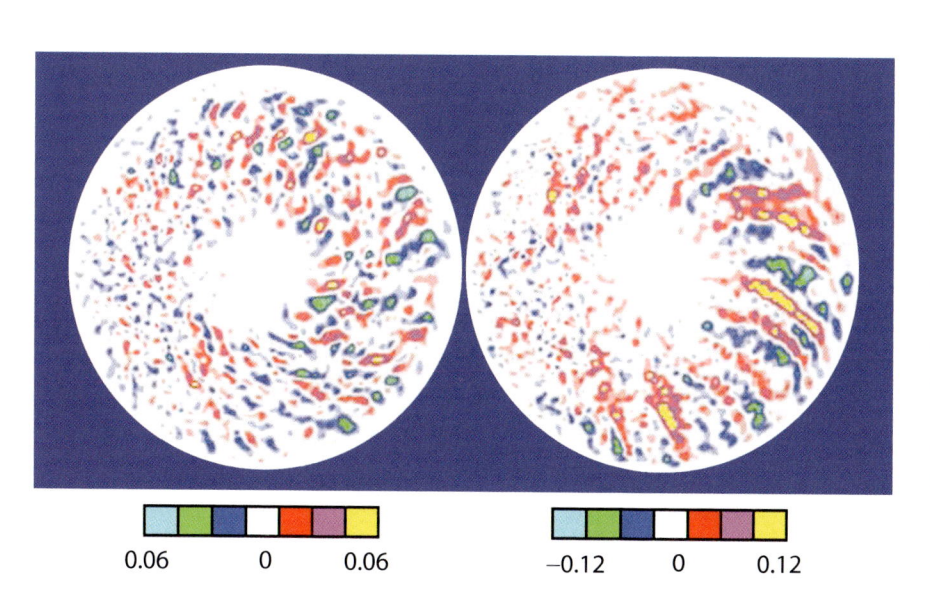

Computer simulations of the effect of sheared plasma flows on turbulence in a tokamak. The image on the right shows electrostatic potential fluctuations in the absence of flows in a turbulent plasma. The turbulence tends to form elongated "eddies" in the radial direction that can transport energy and particles. The image on the left shows how sheared flows can break up these eddies, diminishing transport losses.

Image courtesy of Zhihong Lin, University of California, Irvine, Department of Physics and Astronomy.

Fusion by Light

THERE IS AN ALTERNATIVE MEANS of slamming hydrogen ions together hard enough to fuse: concentrating high-energy laser beams on a small isotope-bearing target. This approach, called *inertial* confinement because the atoms are acted on by the inertia of their own mass, takes one of two forms: direct and indirect drive.

In direct drive, deuterium and tritium are placed in a spherical pellet, typically around the size of an apple seed and made of metal-coated plastic, and the pellet is loaded precisely into the center of a symmetrical array of lasers. The lasers fire simultaneously. The pellet material is vaporized and explodes outward, causing – courtesy of Isaac Newton's third law of motion – an equal and opposite reaction that drives the hydrogen isotopes inward, squashing them together to a density around 100 times that of lead. Then the shock wave produced by the process compresses them even further, heating the plasma to the fusion point. The entire process takes a few billionths of a second.

The indirect-drive method uses basically the same set-up. But the pellet is enclosed in a cylindrical container called a hohlraum (German for "hollow space") made of gold, uranium or some other heavy metal and with the diameter of a dime. When the lasers fire, the metal atoms are transformed into a plasma that gives off X-rays, which in turn strike the pellet surface, causing the same kind of implosion as in direct drive. The deuterium and tritium are forced into a "central hot spot" at which fusion occurs.

The seeming simplicity of inertial confinement (IC) schemes masks a number of daunting technical problems, not least of which is that it is still unclear whether IC fusion can actually produce more energy than it takes to start it. IC ignition depends critically on minutely precise symmetrical placement of all the components – to micrometer tolerances in many cases. That includes the isotopes themselves, which must be deposited in frozen form at a uniform depth on the inside surface of the plastic pellet and maintained at a couple dozen degrees above absolute zero. That

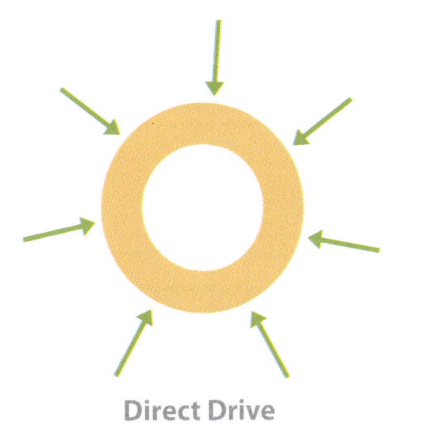

Direct Drive

Indirect Drive

Schematic drawing of direct and indirect drive methods for inertial confinement fusion.

Image revised from Physics 2010 Committee, *Plasma Science: Advancing Knowledge in the National Interest* (Washington, D. C.: The National Academies Press, 2007), p. 86.

temperature constraint leads to questions about how successive unshielded pellets could be injected into a hot chamber, a consideration that may give a slight advantage to indirect-drive schemes, since the hohlraum can provide the pellet some thermal insulation. Then there is the matter of the laser technology. In a working reactor, the lasers would have to fire several times a second for extended periods of time. At present, the most sophisticated facilities can manage a few shots a day between cooling cycles.

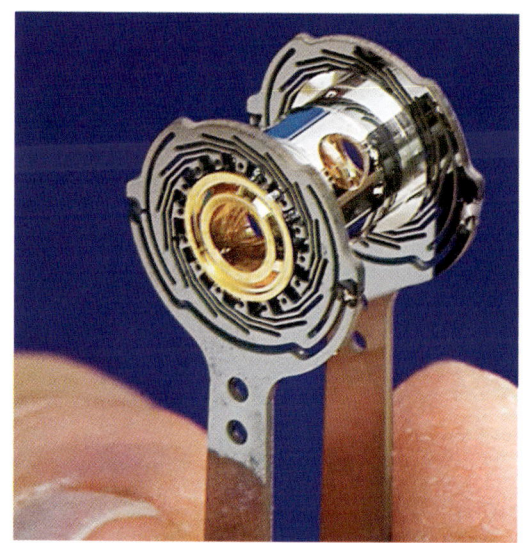

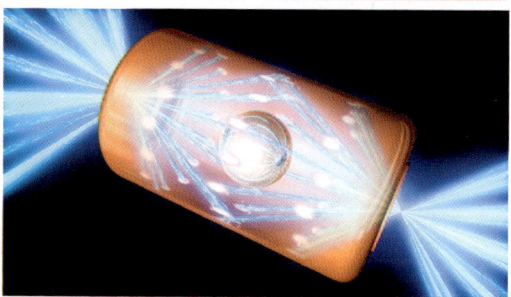

If the obstacles to IC are challenging, the potential rewards are great. Unlike magnetic confinement, where plasmas are relatively diffuse, IC devices can achieve pressures and densities that shed light on the most extreme of astrophysical events, from the supernova death of giant stars to the formation of black holes, stellar evolution, and planetary development. IC experiments can also reveal the behavior of nuclear material under titanic stresses – including conditions engendered by thermonuclear weapons – and a host of unusual hydrodynamic phenomena, such as those that occur when a low-density fluid is driving a heavier one. Patterns and instabilities that arise in those conditions are remarkably similar whether the setting is oil driven into water, a shock system striking an interstellar cloud, or a pellet target vaporized by lasers.

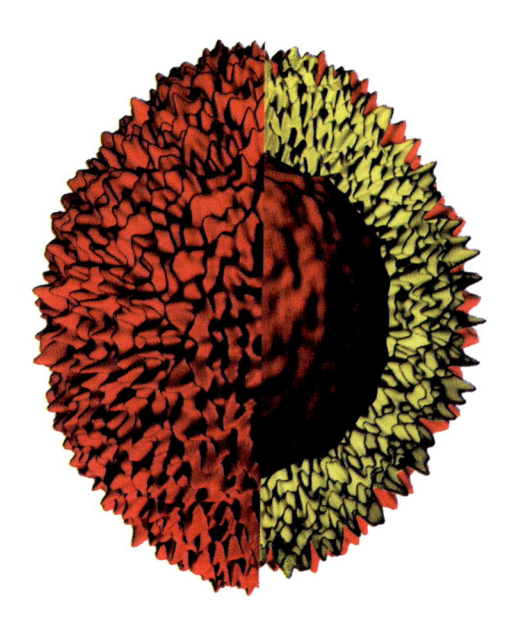

The top frame shows a hohlraum for the National Ignition Facility at Lawrence Livermore National Laboratory. The bottom frame shows an image of the hohlraum serving the dual purpose of holding the spherical target and converting the ultraviolet laser beams to X-rays that are capable of heating the target much more efficiently and rapidly than the laser alone.

Above, the image of a computer simulation performed at the National Ignition Facility using the code HYDRA. It shows the density irregularities produced by fluid instabilities after the laser beam strikes the target.

Images courtesy of Lawrence Livermore National Laboratory, the National Ignition Facility, and the Department of Energy, under whose auspices the work was performed.

The next few years should determine which, if either, of the IC alternatives is suitable for a practical electrical generator. In particular, scientists are looking to the National Ignition Facility (NIF), under construction in Livermore, CA. When completed, it will focus 192 lasers totaling 500 trillion watts of power onto a hohlraum half the size of a spark plug. The ensuing fusion reaction is expected to release ten to 100 times the amount of energy expended to ignite it.

Even if the initial design is unsuccessful, there is an appealing modification that is theoretically feasible: "fast ignition." It employs the same overall technology to compress the hydrogen plasma. But instead of relying entirely on the "central hot spot" to ignite fusion, it fires one final, very short, high-intensity laser pulse into the imploded isotopes, kicking off the reaction. The presumed advantage of this alternative is that it separates the compression and ignition stages, thus reducing the need for ultra-accurate alignment of all components from the start.

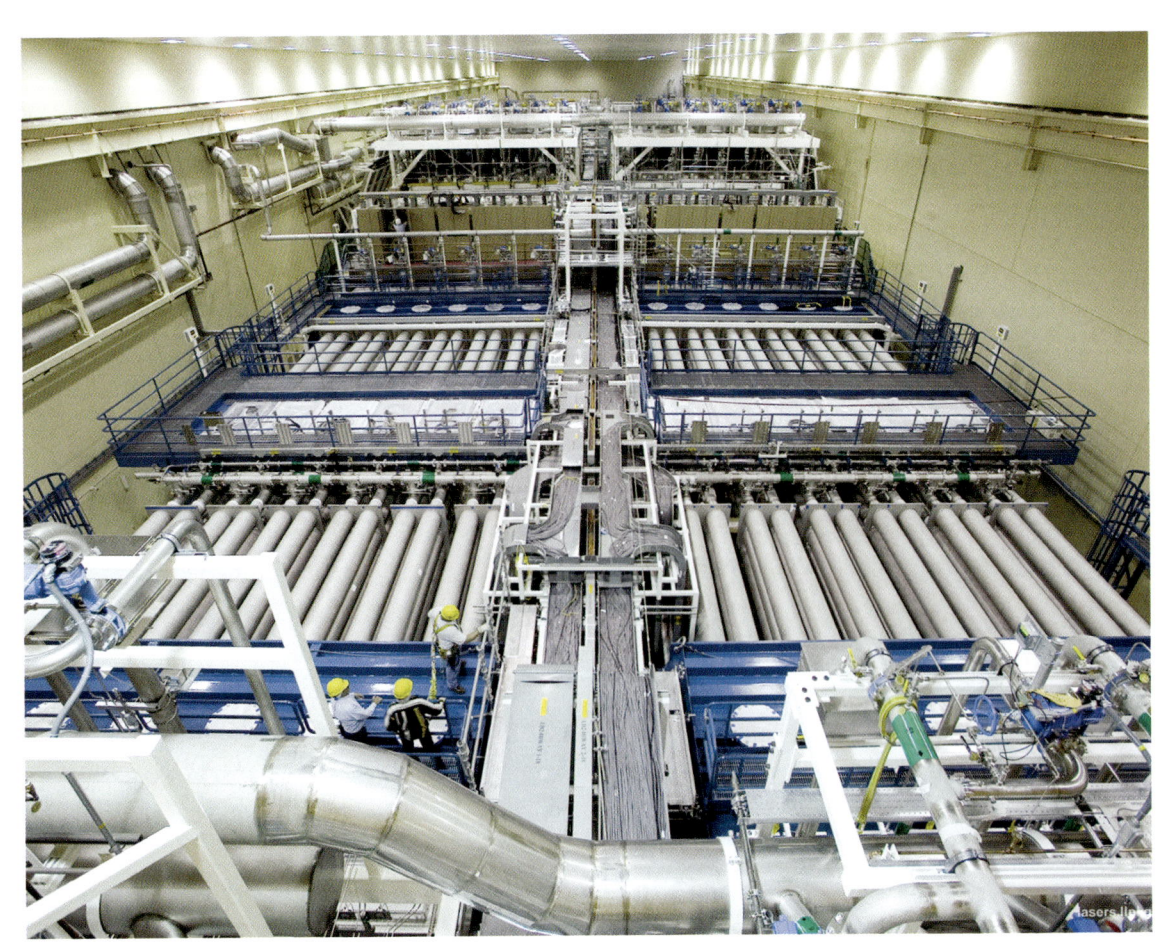

One of the two laser bays of the National Ignition Facility. This bay was commissioned in July, 2007.

Image courtesy of Lawrence Livermore National Laboratory, the National Ignition Facility.

Just a Pinch

> FINALLY, THERE IS A NOVEL VARIANT ON IC that has been around for 60 years, but has attracted renewed interest over the past 20: the "Z-pinch" configuration, so-called because its usual configuration involves current flow in a vertical direction, the z-axis in standard x-y-z three-dimensional coordinates. Z-pinch devices use extremely large currents to vaporize thin metal wires arranged in a birdcage-like design with a hohlraum in the center. The metal atoms turn into a cylindrical sheet of plasma as the moving current induces a magnetic field. The interplay of current and field produce a force (the Lorentz force) that acts perpendicular to the current flow, squeezing everything toward the hohlraum at the center of the cage.

Z-pinch machines – and a related "X-pinch" design that uses crossed wires – are usually employed to generate brief bursts of X-rays for scientific experiments. But the mighty 120-foot-wide apparatus at Sandia National Laboratory in New Mexico, which rams tens of millions of amperes through its wires in less than a tenth of a millionth of a second, has proven capable of fusing deuterium (see the image on page 14).

Whether or not it becomes a viable contender for commercial fusion energy, Z-pinch science, along with accelerating progress in IC and magnetic-confinement systems, is fast expanding the frontier of an increasingly exciting field: high-energy-density physics (HEDP). Less a specific discipline than a constellation of interrelated and synergistic research subjects, HEDP's investigations of matter in extreme conditions are bringing the perspective of newly possible laboratory experiments in plasmas, particle beams, and lasers to the understanding of phenomena from condensed matter, materials research, fluid dynamics, and nuclear science to the most violent events of astrophysics, including relativistic jets blasted billions of miles into space by black holes.

Indeed, fusion research in general serves numerous high-priority national goals. It expands scientific understanding of hydrodynamics, of chaotic, turbulent, and "emergent" (self-forming) patterns, and of geomagnetic phenomena. It contributes invaluable insights to other sciences, such as meteorology, applied mathematics, and more. It produces a human inventory of scientific talent to keep the United States at the forefront of many disciplines, including the global quest for non-carbon energy sources. And, as Chapter 5 demonstrates, it explains the furious forces at play in our plasma universe.

Marshall Rosenbluth
(1927–2003)

PLASMA PHYSICIST

NATIONAL MEDAL OF SCIENCE, 1997

MAXWELL PRIZE IN PLASMA PHYSICS, 1976

Marshall Rosenbluth

SO MANY SCIENTISTS have contributed to the field of nuclear fusion that the major names alone make a long and distinguished list. But even in that select group, plasma theorist Marshall Rosenbluth stands out. The New York native earned his doctorate from the University of Chicago at the age of 22 – despite two years' service in the US Navy! – and was promptly recruited by his advisor, Edward Teller ("the father of the hydrogen bomb"), to work at Los Alamos National Laboratory.

Rosenbluth, who witnessed the first thermonuclear explosion in the Pacific, remained at Los Alamos until 1956, when he joined a little-known California start-up company called General Atomics, which would become a world leader in fission and fusion technology. Shortly thereafter he obtained an appointment at the University of California at San Diego, and went on to positions at Princeton, the University of Texas, and other institutions. Along the way, he had a hand in virtually every major development in both magnetic and inertial confinement fusion.

In 1965, he organized a meeting between American, European, and Russian plasma physicists in Trieste, Italy, opening the door to international collaborations in fusion. He was a member of the National Academy of Sciences, received the National Medal of Science from President Clinton for service to his field and his country, and accumulated numerous other honors and distinctions. But perhaps none was more admired by his peers than the nickname he earned for his profound understanding: "the pope of plasma physics."

THE COSMIC PLASMA THEATER:
Galaxies, Stars, and Accretion Discs

IF THERE WERE A *Guinness Book of Cosmic Records*, listing the most astonishing, prominent, perplexing or destructive astrophysical events in the known universe, plasmas would appear, directly or indirectly, in nearly every entry.

That shouldn't really be surprising. According to the Big Bang theory, the cosmos erupted into existence about 13.7 billion years ago and spent most of its first 300 000 years as unalloyed expanding plasma until it cooled sufficiently for the first neutral atoms to form.

And even today, intact atoms make up no more than a tiny fraction of the observable universe. As a result, the effort to understand a broad spectrum of phenomena – from the way stars and planets coalesce out of plasma discs, to the evolution of galaxies, to the magnetic maelstroms around black holes and the firing of interstellar bullets we call "cosmic rays" – requires the perspective of plasma science.

What *is* surprising, however, is the remarkable ubiquity of magnetic fields in the universe, and the relationship of those fields to plasmas on size and intensity scales that range over dozens of orders of magnitude. There is no obvious reason that so many objects and aggregations of objects – including individual stars and planets, the interstellar medium, whole galaxies, and even the vast evacuated spans between galaxies – should have associated long-lived magnetic fields. But they do, and the dynamics of these bodies are strongly affected by the interplay of plasmas and fields.

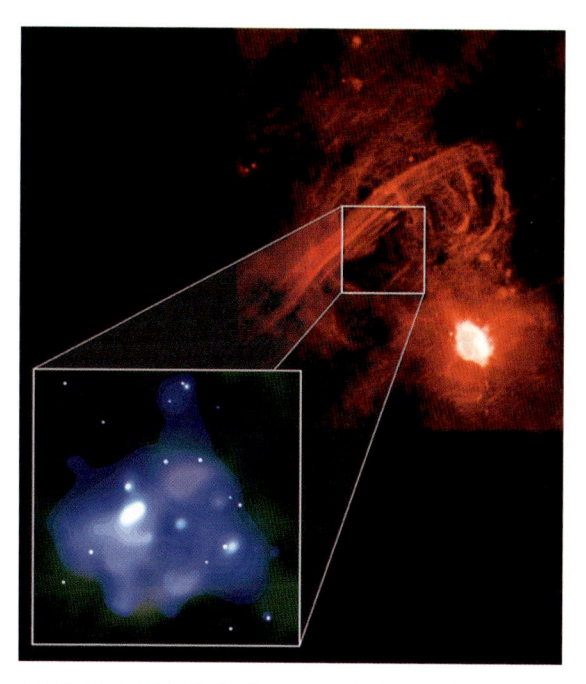

A region called "The Arches" near our galactic core shows structures formed by intense magnetic fields around the central black hole. X-ray image (blue) courtesy of NASA/CXC/SwRI and Gladstone *et al.*; optical image (red) courtesy of NRAO/AUI/NSF.

Shaping Up

ASTROPHYSICAL OBJECTS form as clouds of matter are drawn together by gravity, collapsing at accelerating rates. The converging plasma particles, of course, do not have an isotropic (perfectly uniform in all directions) distribution of density and velocity. Over time, those tiny irregularities start to add up. Many, but not all, cancel each other out. The growing gravitationally bound object ends up with a net collective motion, such as the axial spin of stars like the Sun, or the stately rotation of spiral galaxies such as the Milky Way.

That motion in itself, however, is not necessarily sufficient to generate a large overall magnetic field. If the effect of each electron were exactly offset by the opposite effect of a proton, the total effects would sum to zero. But given the immense numbers of particles involved, even the very slightest imbalance in the collective motions of electrons and protons can constitute a significant net current, and hence an accompanying magnetic field.

So it's not hard to see how random outcomes might have started an organized "seed" field started at various venues in the nascent universe, or how they generate them in bodies forming today. But what is much harder to understand is how these fields have persisted for so long – billions of years in the case of the Earth and the Sun, and even longer for galaxies. For that to happen, the initial field has to be reinforced continuously in some way that keeps it from simply dissipating as it encounters resistance from matter.

Electrical machines designed to generate self-sustained magnetic fields and electric currents, for example to charge a car's battery, are traditionally called dynamos. Physics has appropriated this word to signify naturally occurring systems that do the same thing. An electrically conducting medium moves through a magnetic field, inducing currents, which in turn generate their own magnetic field that reinforces and amplifies the original field.

In the local case of the Earth, the conductor is provided by the molten iron and nickel surrounding the planet's super-hot inner core. Heated by the core, the metal moves in two ways at once: rising and falling in convection loops; and revolving sideways as the planet spins. This system, scientists surmise, produces a dynamo effect as the metal passes through magnetic fields. Similar convective and rotational motion occurs in the Sun. (To make matters even more complicated, both bodies reverse their polarities at intervals – about every 11 years in the Sun and every few hundred thousand years in the Earth. Remarkably, dynamo theory can account for these periodic reversals.)

Galaxies, of course, do not exhibit convection. But as they form, they attract intergalactic plasmas with embedded magnetic field lines that are entrained with the plasma mass. As the galaxy takes shape and evolves, there are numerous motions that can twist and magnify existing magnetic field lines – and generate new configurations – that together form patterns allowing large, self-perpetuating collective fields to form. Although the net field in our galaxy may seem weak (about one-millionth of a gauss, compared to about half a gauss for the Earth's field at the surface and one to two gauss on the surface of the Sun), its effect on galactic matter is nonetheless large.

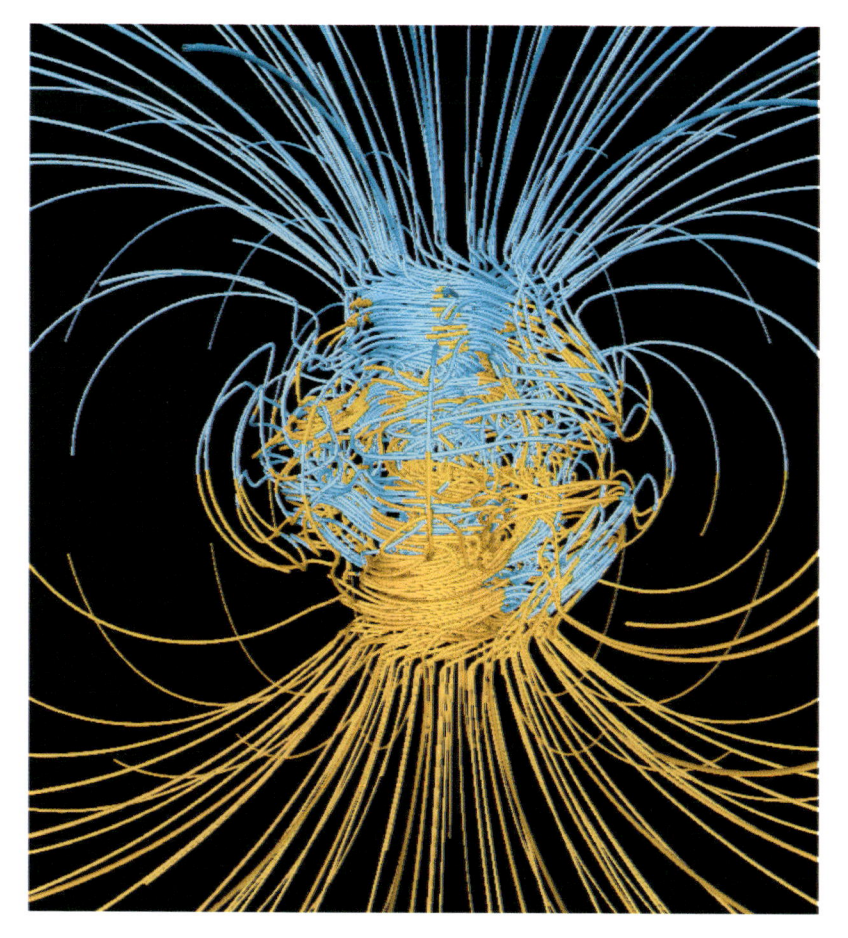

A snapshot from a computer simulation of the generation of the Earth's magnetic field (the geodynamo). The 3-D field is illustrated with magnetic lines of force, which are colored blue where the field is directed inward and gold where it is directed outward. The rotation axis of the modeled Earth is vertical in the image. (The field is presumably much more intense and complicated inside the Earth's fluid core where it is generated.) Outside the core the field has a dominantly dipolar pattern. Several spontaneous reversals of the magnetic polarity have occurred during the simulations, similar to those seen in the Earth's paleomagnetic record. These geodynamo simulations now span more than a million years, using an average computational time step of about 15 days.

Image courtesy of Gary A. Glatzmaier, University of California, Santa Cruz, and Paul H. Roberts, University of California, Los Angeles.

Going to Extremes

> THERE ARE, HOWEVER, a number of astrophysical plasma phenomena that produce objects with magnetic fields of gargantuan power. One class of such objects is formed when a massive star explodes because, as it runs out of fusion fuel, the radiation pressure pushing outward no longer balances the gravitational force pulling inward. Gravity suddenly and catastrophically prevails. The star implodes and then blasts outward to produce a supernova, and the far-flung debris typically forms a gassy cloud called a nebula. But in the course of compression, supernovas may also create a super-dense body called a neutron star.

As recently as 40 years ago, the existence of neutron stars was regarded as a somewhat fanciful supposition. Now we know them to be familiar features of the universe, but they nevertheless continue to defy imagination.

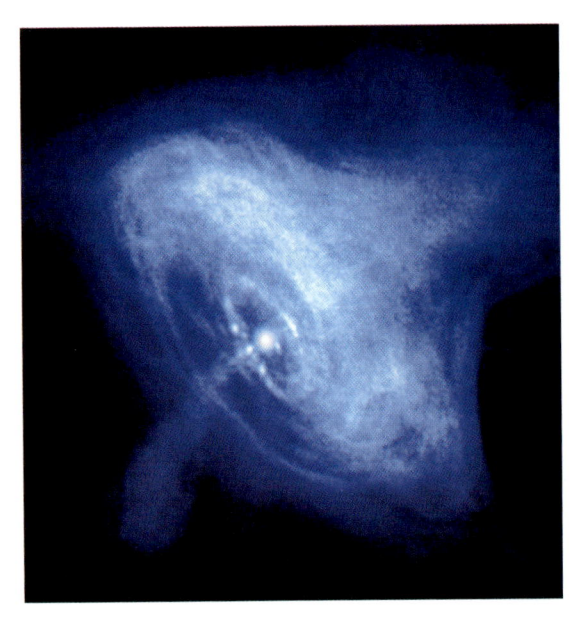

The Crab Nebula is the remnant of a supernova explosion that was seen on Earth in 1054 AD. It is 6000 light years from Earth. At the center of the bright nebula is a rapidly spinning neutron star, or pulsar, that emits pulses of radiation 30 times a second.
Image courtesy of NASA/CXC/ASU and J. Hester *et al.*

Each was once the center of a collapsing supernova star, so crushingly compacted that the protons and electrons in the original core plasma were squeezed together to form neutrons. In effect, the entire object has a density equivalent to an atomic nucleus, resulting in a mass several times that of the Sun contained in a sphere only a couple dozen kilometers in diameter – about the size of Kansas City – at something like 4×10^{17} kg m^{-3}. That's dense: a teaspoon of such exotic matter would weigh about 100 million tons on Earth.

But mass is merely one of a neutron star's outlandish properties. Another is the enormous magnetic field strength produced when the size of the original star is abruptly reduced by a factor of a billion or so, cramming field lines together. Yet another is the dizzying spin rate. As anyone who has watched a skating competition knows, a pirouetting skater's rotation rate speeds up as she pulls in her arms and legs. This same effect, the result of conservation of angular momentum, also occurs as stellar cores shrink into neutron stars, raising their spin from a leisurely pace (the Sun, for example, takes a full month to make one revolution) to as much as 1000 revolutions per second.

All those attributes combine into a mighty cosmic beacon. On the neutron star's surface, electrons and protons that didn't quite get squashed into neutrons are picked up by huge electrical forces and propelled along the magnetic field lines. That motion, in turn, generates a beam of electromagnetic radiation that flies outward from the star's north and south magnetic poles. If the magnetic poles are tilted slightly from the neutron star's axis of rotation, these radiation beams will propagate in a cone-shaped pattern. So a far-away object, such as a telescope on Earth, will be struck by the radiation only once every revolution, and the beam will seem to be blinking, much as lighthouse rays appear to a ship at sea.

Radio astronomers first detected such a strange radiation pattern in 1967. Because of its intermittent signal, it was dubbed a "pulsar." Since then, numerous variations have been discovered. Some emit X-rays, and a rare few are capable of radiating "soft" (lower-energy) gamma rays. Gamma rays are the most energetic form of electromagnetic radiation, and it takes such a powerful class of neutron stars to generate them that the objects have been given their own name: magnetars. Magnetars are presumed to be neutron stars that form with an extraordinarily high spin rate, endowing them with magnetic field strengths in the range of a thousand trillion gauss or more. The gamma radiation from these tiny power-balls, the largest of which is smaller than Rhode Island, is detectable from hundreds of thousands of light years away – far outside our galaxy.

(1) In this depiction, the core of a massive star collapses, ending its life – though there is little effect visible at the surface. Deep inside, twin beams of matter and energy begin to blast their way outward.

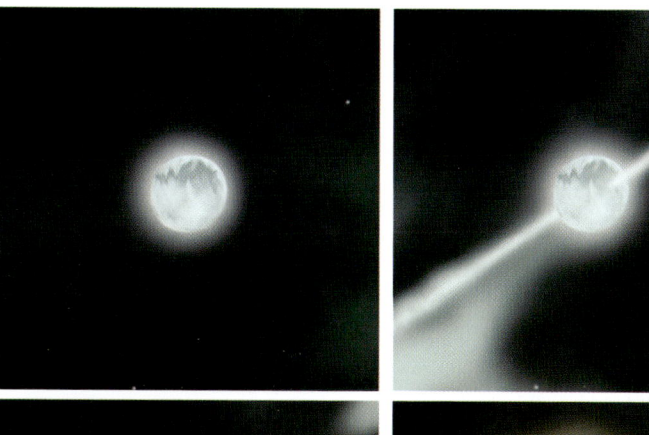

(2) Within seconds the beams have eaten their way out of the star, and observers on Earth see it as a gamma-ray burst, GRB 060729A.

(3) The outer envelope of the star explodes outward, causing a supernova.

(4) At the heart of this event, the core has shrunk into a dense magnetar, a neutron star possessing a magnetic field trillions or even quadrillions of times stronger than Earth's. The magnetism is what powers the long glow of X-rays seen by Earthbound scientists.

Images courtesy of NASA E/PO, Sonoma State University, Aurore Simonnet.

Discs and Holes

UNBELIEVABLE AS IT MAY SEEM, magnetars are by no means the strongest sources of plasma-related radiation in the universe. The highest energies ever observed belong to brief gamma-ray bursts apparently engendered by the collapse of a massive star into a black hole, or by the collision of two neutron stars orbiting one another in a binary system. For sheer size, however, the cosmic title is held by the doomed whirlpools of plasma surrounding supermassive black holes that lie at the center of most (or perhaps all) large galaxies, swallowing up matter in an insatiable binge of astrophysical conspicuous consumption.

Black holes can form in several ways, and some may be left over from the Big Bang era. But most of those identifiable from Earth are thought to result from the gravitational collapse of giant stars with masses a million, or even a billion, times larger than the Sun's. Their distinctive hallmark is utter darkness: the hole's gravitational field is so strong, and the corresponding warp of space-time so severe, that not even light itself can escape once it passes the no-return boundary called the event horizon. Consequently, black holes are, by definition, invisible. We observe them only through their cataclysmic effects on surrounding plasma. But that still leaves plenty to see.

As surrounding matter spirals inward toward the hole, it forms a structure called an accretion disc, somewhat resembling a donut, but with a teardrop-shaped cross-section. Accretion discs are a common sight in the cosmos wherever diffuse matter orbits a central gravitational source. Indeed, our own solar system likely formed from a circumstellar accretion disc, and similar formations accompany young stars, white dwarfs, and neutron stars. But the accretion discs around galactic black holes – which are assumed to be the source of radiation-rich "active galactic nuclei" visible in a wide range of wavelengths – are in a class of their own.

As the plasma in the disc whirls toward oblivion, it is subject to numerous physical processes, occurring simultaneously, that produce telltale effects. One, of course, is frictional heating as matter is compressed toward the innermost edge of the disc just before the event horizon. Compression can raise temperatures to a billion degrees and produce strong X-rays. Meanwhile, the incessant inflow of plasma releases gravitational potential energy – the energy the matter had by virtue of its distance from the hole's center of mass. This process seems to be one of the most efficient means in nature to convert matter into energy. By one estimate, it is 50 times more efficient than nuclear fusion!

But very little is known conclusively about the origin, specific fine structure, and magnetic dynamics of discs, and many essential problems remain to be solved. For example, it seems clear that discs do not behave in the conventional orderly fashion described by Johannes Kepler in the seventeenth century to explain the rotation of planets around the Sun. In that model, rotational velocity becomes proportionally faster the closer an object is to the inner part of the disc, allowing even close-in objects to remain in orbit. But in an accretion disc, matter in the innermost disc slows down enough to plunge over the event horizon.

That raises a major question. It is a fundamental law of physics that, in all physical systems, momentum is conserved; that is, there is always the same amount of it before and after something happens. But in the case of the matter spinning in an accretion disc, why does matter lose momentum at the disc's inner edge? And how does it affect the behavior of the disc?

The giant elliptical galaxy NGC 4261, viewed as a composite of ground-based optical and radio wave observations. (Left) Photographed in visible light (white), the galaxy appears as a fuzzy disc of hundreds of billions of stars. A radio image (orange) shows a pair of jets emanating from the core. (Right) Hubble Space Telescope image of the core of the galaxy, which is presumed to harbor a black hole. The dark inner disc represents a cold region, which is surrounded by a very hot accretion disc located within a few hundred million miles of the black hole.

Image courtesy of National Radio Astronomy Observatory, California Institute of Technology; Walter Jaffe/Leiden Observatory, Holland Ford/JHU/STScI, and NASA.

In recent years, astrophysicists have devised an ingenious explanation, called magnetorotational instability, which has become widely accepted. The plasma circling a black hole passes through the hole's poloidal magnetic field, and in turn creates a toroidal field as it grabs and stretches field lines. In the resulting arrangement, radially adjacent regions of the disc are linked by a magnetic connection that acts like a rubber band or spring. Consider two neighboring groups of particles, one of which is farther outward from the center. The inner group is naturally disposed to move faster than the outer one. But the magnetic connection between them retards the inner group as if it were being pulled backwards.

Simultaneously, the same tension tugs the outside group forward, making it move faster than it ordinarily would. Thus angular momentum is transferred outward in the disc, producing irregularities and turbulence, but also allowing the innermost matter to slow down and drop into the hole.

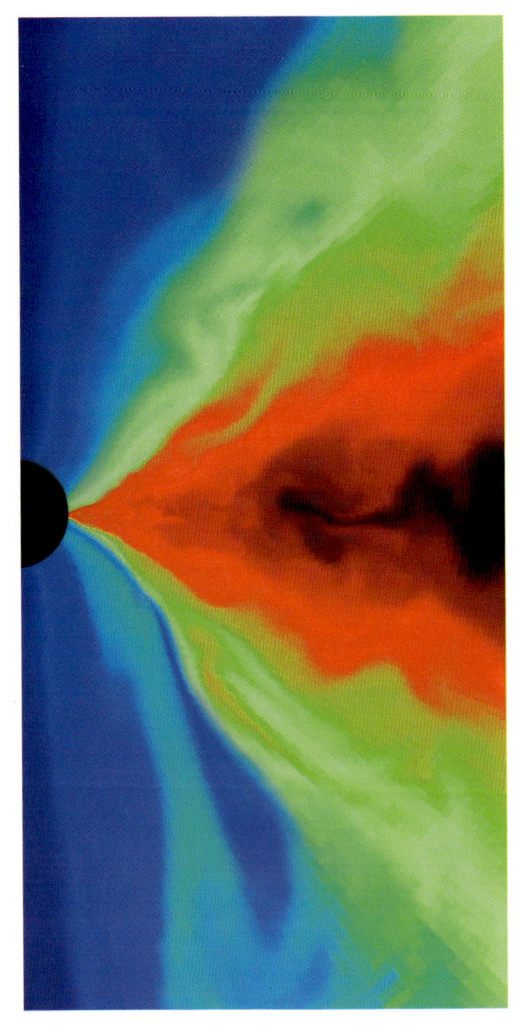

This snapshot from a computer simulation of density variations in the matter surrounding a black hole accretion disc reveals a remarkable amount of complex structure and turbulence.

Image courtesy of J. Stone, Princeton University.

Jet Propulsion

SCIENTISTS HAVE MADE less definitive progress in explaining one of the most spectacular phenomena in the universe: the "jets" of particles and radiation that stream off from the centers of accretion discs and can extend for huge distances into surrounding space. In fact, jets are the longest coherent structures observed in the universe. The jet from a nearby galaxy called M87, for example, stretches for about 600 000 light years, six times the length of our own Milky Way galaxy at its widest point. And recently radio astronomers found a jet one million light years long.

Particles in the strongest jets can travel at a substantial fraction of the speed of light and remain collimated (held in a tight, narrow beam) over long distances. Although some jets are asymmetric, or "one-sided," the prototypical pattern is bipolar, with approximately equal streams in exactly opposite directions.

It seems clear that jets are produced by the joint action of large-scale magnetic fields and the motion of matter in an accretion disc surrounding a black hole or neutron star. But precisely how they interact is still a mystery. Most theories assume that the magnetic poles of the central object, such as a spinning black hole, are involved in "aiming" the jets.

One model describes that region, which extends perpendicular to the accretion disc plane, as a sort of tornado-like vortex or funnel. Centrifugal force would ordinarily keep the funnel's center relatively empty of particles, but full of spiraling magnetic field lines. However, if there is some mechanism that would boost plasma up into the walls of the funnel, the particles would be constrained to follow the field lines. And, conveniently, their helical motion while moving outward would create a toroidal magnetic field that would serve to squeeze the beam and keep it collimated.

It is difficult, however, to determine what sort of force would propel plasma off the accretion disc and up into the jet. One possibility is that, if the plasma pressure at the base of the funnel were somehow high enough, the much lower pressure in the funnel would be sufficient to draw particles from the disc's "corona" – a loose, diffuse cloud that surrounds the main mass of the disc – upward into the wall.

An image of a jet from the M87 galaxy.
Image courtesy of NASA.

In the Firing Line

THE SAME KIND OF RESEARCH may help plasma scientists understand the source of another astrophysical enigma: cosmic rays.

That name may be somewhat misleading. Cosmic rays are not rays in the conventional sense, but individual matter particles that strike the Earth's atmosphere and smash into air molecules. About 90 percent of them are single protons, another 9 percent are helium nuclei, and the remaining 1 percent comprises electrons or (rarely) miscellaneous atomic nuclei. Many of the "cosmic" rays arriving at our planet are not cosmic in origin, but are blown in from the Sun. They are more properly called solar energetic particles, and are propelled by coronal mass ejections or solar flares.

It is possible to determine a cosmic ray's energy content directly in instruments placed well above the bulk of the atmosphere. But it can also be measured indirectly from the shower of particles that results when a cosmic ray particle collides with oxygen and nitrogen molecules in the air. Either way, the results are baffling. Cosmic rays vary in energy over more than 12 orders of magnitude, reaching nearly 10^{20} eV. (Moreover, there does not seem to be any upper limit.) That is so much energy that even physicists who study the most cataclysmal processes in the entire universe cannot explain them.

To get a sense of scale, the Large Hadron Collider, the most powerful particle accelerator ever built, shoots streams of protons around a 27-km ring at 99.9 percent the speed of light. That gives the particles energies of a few trillion eV. But the most energetic cosmic rays observed to date have 100 million times that much energy. Where could they possibly come from?

Trying to answer that question has, so far, been largely a matter of ruling out most possibilities without ruling any in. Below 10^{15} eV, cosmic rays might have obtained their kicks from various sources within our galaxy. Above that level, abnormally energetic supernovae with relativistic jets or pulsars with magnetic fields in excess of a trillion gauss could conceivably do the job. But if that were the case, then the cosmic ray flux would tend to come only from certain preferred directions correlated with galactic structure. But that is not the case. In fact, the spatial distribution of the highest-energy cosmic rays (above 10^9 eV) is completely uniform; they come from every direction with equal probability.

Outside the galaxy, potential sources are still limited. Physicists assume that a cosmic ray particle is unlikely to travel more than about 100 megaparsecs (around 300 million light years) without running into something or having its energy attenuated in some way. Within that distance from our galaxy, there are several options under consideration.

Fluxes of Cosmic Rays

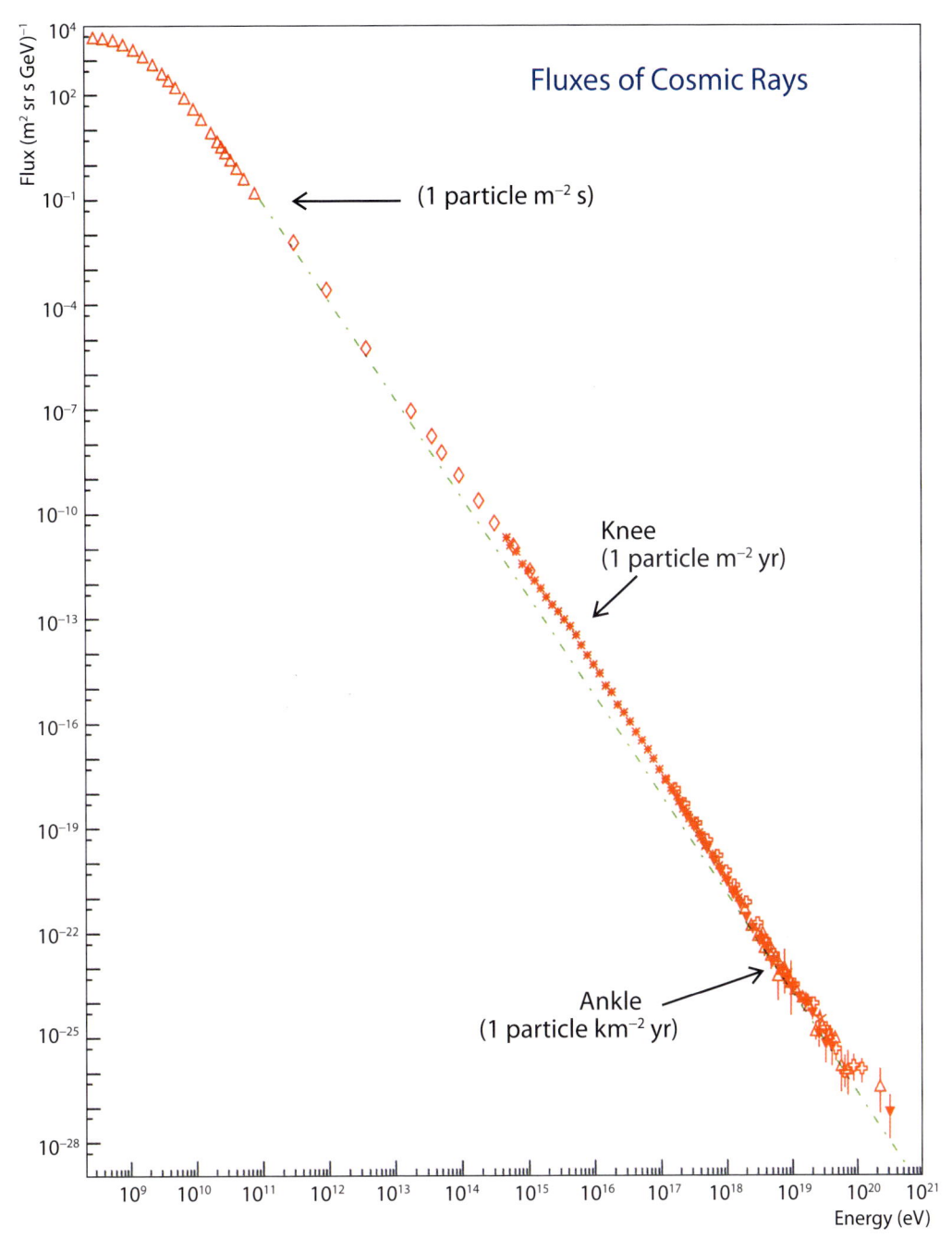

(1 particle m^{-2} s)

Knee
(1 particle m^{-2} yr)

Ankle
(1 particle km^{-2} yr)

This seemingly simple plot tells a complicated story about the behavior of plasmas across the universe. It shows how cosmic rays detected at the Earth vary over an amazing 12 orders of magnitude in energy – from 100 million eV to 100 million trillion eV. The vertical axis shows the total number of cosmic rays arriving per unit area, with the most common at the top. The horizontal axis shows energy, with the highest to the right. Those in the "ankle" area of the plot originate far outside our galaxy, propelled by processes that are still not understood.
Image courtesy of Simon Swordy, University of Chicago.

THE COSMIC PLASMA THEATER

There is, for example, some evidence that arriving rays have trajectories that correspond to the known locations of active galactic nuclei. But it appears unlikely that a particle in an active galactic nucleus could attain suitable energy levels. And, even if it could, a cosmic ray would have to fight its way out of the thicket of fields and plasmas in the galactic nucleus with a lot of energy intact. Gamma-ray bursters are another presumptive source. But they appear to manifest the requisite energies only once every 100 years or so, and there simply aren't enough of them within the 100-megaparsec limit to account for the observed rate of super-high-energy cosmic rays arriving at Earth, which is about one per year.

Finally, there are several kinds of plasma events that could amplify a particle's energy to the highest observed levels if it already had a good start. One involves shock acceleration, in which an already fast-moving particle gets a surge of energy by riding the shock-wave from an exploding supernova or equivalent occurrence.

Another is a fascinating class of actions, called Fermi acceleration after the celebrated physicist who theorized it to explain high-energy cosmic rays.

Enrico Fermi speculated that if a particle encountered a fast-moving interstellar magnetic field (dubbed a "magnetic mirror") in a head-on fashion, it would bounce off it with increased energy, much like a ping pong ball hitting a moving paddle. This is a very promising hypothesis because the energy transferred to the particle would be proportional to the *square* of the moving mirror's velocity. In addition, there are arrangements in which a particle could actually bounce back and forth between two magnetic mirrors, one ahead of and one behind an advancing shock wave.

Plasma astrophysicists will continue to investigate these and other exceptional processes as they strive to describe the most forceful actions of which nature is capable. Meanwhile, to the extreme good fortune of everyday society, their colleagues are perfecting ways to harness the power of plasmas on the mundane energy scales in which human beings abide.

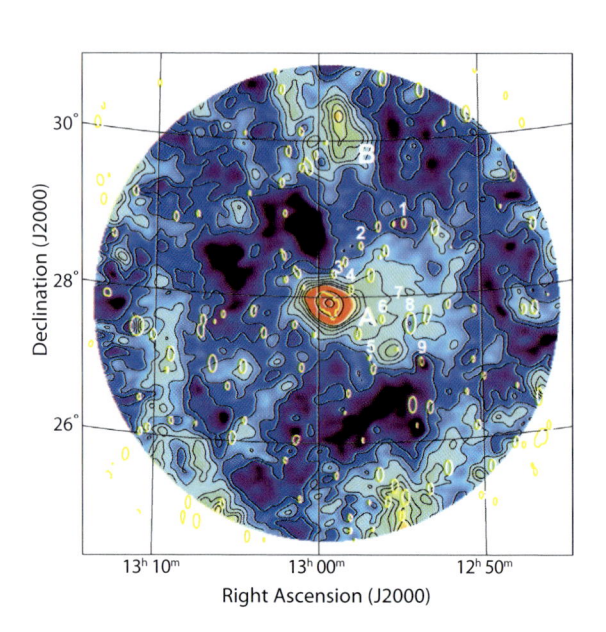

This image, produced by radio telescopes in Puerto Rico and Canada, shows an enormous cloud of plasma (light blue area to right of center) about 80 times the width of our Milky Way galaxy. The cloud is 300 million light years away in a region of space occupied by the Coma Cluster of galaxies (red in the image). The source of its energy is not known with certainty. Energetic particles escaping from such clouds could potentially make their way to Earth as cosmic rays of extremely high energy.

Image courtesy of Philipp Kronberg et al., Los Alamos National Laboratory / Arecibo Observatory / DRAO. Reprinted with permission of the AAS: *The Astrophysical Journal* **659** (2007), p. 269; copyright 2007.

Putting Plasmas to Work

NO MATTER WHERE we live or work in the modern world, plasmas pervade our lives. They are seldom seen directly. But they are essential to microelectronics and telecommunications, material coatings and surface treatments, creation of bio-inert devices for surgical implantation, key processes in the textile and clothing industries, and, of course, light fixtures of numerous kinds.

All of these applications, and many more, are subsumed under the general designation of "low-temperature" plasma science, although "low," as we shall see, is a relative term. At energies below 10 eV (100 000 K), plasmas can interact in myriad usable ways with their surroundings, depositing layers of material on surfaces, prompting chemical reactions, irradiating nearby objects, and even providing propulsive power.

To get a sense of the broad utility of the low-temperature plasma domain, one need look no farther than the ceiling above, where there is likely to be a fluorescent lighting fixture – the single most familiar plasma appliance in modern society. Fully one-tenth of all the electrical power transmitted in the United States is employed in producing plasmas in such lamps. Without their dependable efficiency, which is about five times that of incandescent lights, America's annual energy consumption would be drastically higher and its conquest of darkness far less complete.

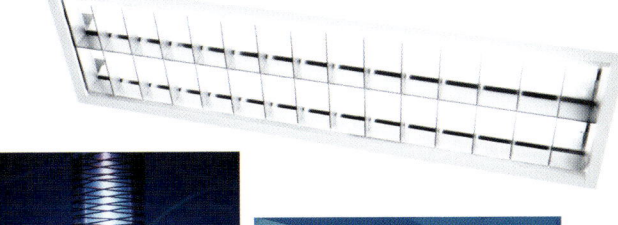

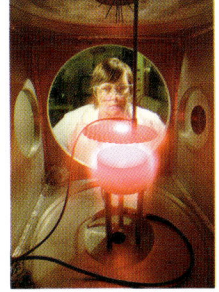

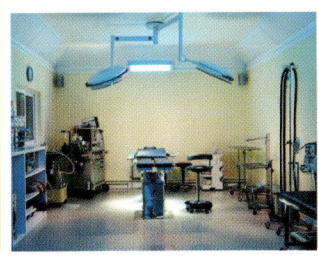

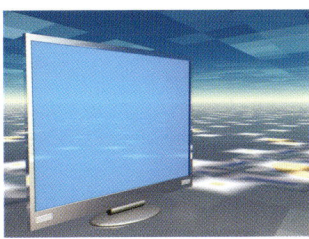

A FLUORESCENT LIGHTING FIXTURE — THE MOST FAMILIAR PLASMA APPLIANCE IN MODERN SOCIETY.

A Plasma to Read By

THE FLUORESCENT LIGHT has evolved into a compact marvel of ingenuity since the first commercial prototypes appeared in the 1930s. Within its tube is an inert gas, such as argon, and a tiny quantity of mercury. When the lamp is switched on, electrodes at each end of the tube emit a current of energetic electrons, ionizing the mercury and creating the plasma.

The plasma electrons become surprisingly hot: some reach 11 000 K, several times the temperature of the tungsten filament in an incandescent lamp, and far hotter than the surface of the Sun. So instead of emitting visible light, they radiate chiefly in the ultraviolet (UV) range. (A fluorescent tube, however, remains cool enough to touch at full power. The reason is that, while many of the plasma electrons are furiously energized, the more massive ions and neutral atoms in the glass are not.)

The human eye is not equipped to detect UV light, so a further step is necessary to make a fluorescent lamp truly illuminating. The inside surface of the glass tube is coated with a thin layer of phosphors, minerals whose molecules perform a neat and useful trick: they absorb high-energy UV photons, and emit visible light. The chemical composition of the phosphors can be adjusted to make the light slightly "warmer" or "cooler" as the intended use requires.

One final aspect of the design is worth noting, because without it fluorescent lights would not work at all. As explained in Chapter 1, the electrical conductivity of a plasma decreases as its temperature rises. So running a stream of electrons through the ionized gas in the tube would soon produce such a low resistance that the current would spike and burn out the unit or trip a circuit breaker. The problem can be solved by using a simple resistor, but that results in energy wasted as heat.

A cunning solution to both problems is to add a coil called a magnetic ballast (or, often nowadays, its transistorized equivalent). As the plasma heats up and the current gets stronger, the ballast stores that energy

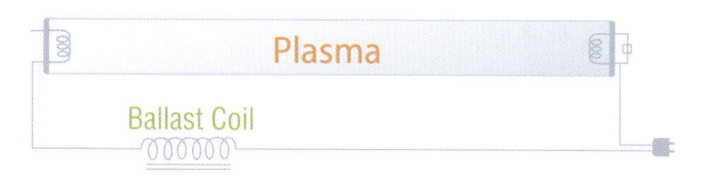

The mercury plasma in a fluorescent light can react so fast that 60 Hz alternating current could shut it down. A ballast coil solves the problem.

Image adapted with permission from David P. Stern.

Part of this fluorescent tube has no inner phosphor coating, revealing the bright blue light from the plasma.

Image courtesy of Paul Kevin Picone/PI Corp and OSRAM SYLVANIA Inc.

SO BRIGHT ARE THESE LIGHTS THAT THEY MAKE PLASMA THE DISTINCTIVE VISUAL HALLMARK OF OUR PLANET.

as an increase in its magnetic field. That process slightly retards the rise in current. Then, when the current falls in the course of reversing direction, the release of stored magnetic energy prompts a compensating voltage surge overcomng the problem caused by the drop in resistance of the plasma.

Fluorescent bulbs are bright enough for home, office, and commercial use. But for the higher luminosity needed in large areas, such as highways and athletic fields, researchers have devised lamps in which the light comes directly from the plasma, without the need for fluorescence. In these lamps, the higher pressure of the gas ensures that the direct plasma emissions are an efficient and appropriately colored source. Such "high-intensity discharge" units employ current arcs flowing through a tube enclosing an inert gas, some sodium or mercury, and, often, a quantity of various forms of halogen compounds – that is, those containing fluorine, chlorine, iodine, or bromine. So bright are these lights that they are the only human artifacts readily seen from space, making plasmas the distinctive visual hallmark of our inhabited planet.

Seen from space at night, the United States is aglow from sea to sea, thanks in large part to plasma lights.

Image courtesy of Craig Mayhew and Robert Simmon, NASA-GSFC, based on Defense Meteorological Satellite Program data.

Walls of Light

PLASMAS ALSO PLAY an important role in display technologies, most notably in flat-screen TVs and video monitors. Although these devices may appear different from their fluorescent-light cousins, the operating principles are basically the same: a plasma screen consists of several million tiny fluorescent cells, each typically about the size of the period at the end of this sentence, arranged in a dense two-dimensional grid and sandwiched between two glass plates.

On either side of the cells is an array of electrodes that lie between the cells and the glass plates. On one side are hundreds of parallel rows of transparent electrodes running vertically; on the other are hundreds that run horizontally. Each plasma cell is situated at an intersection of a vertical and horizontal electrode. (The total number of electrodes determines the "resolution." So a high-definition screen with 1920 horizontal electrode rows and 1080 vertical rows has a total resolution of 1920 × 1080: about 2 million dots.)

The plasma cells are arranged in groups of three (one each for red, green, and blue light) for each picture element, or "pixel," and every cell contains a minuscule quantity of noble gas, such as xenon and neon. If both electrodes are activated, above and below the cell, the voltage is high enough to overcome a built-in resistance layer. A current flows through the gas, and the resulting plasma gives off UV radiation. The UV photons then strike phosphors on the cell walls, where they are absorbed and re-emitted as red, green, or blue light, depending on the specific chemical composition of the phosphor material. If three adjacent cells produce equal maximum illumination, the resulting light appears as a white dot. If none is illuminated, that spot appears black. Between those extremes are as many as 16.7 million possible color variations per pixel, determined by how long the plasma is excited in each color cell, or "subpixel."

Designing such screens demands sophisticated plasma science because of the exquisite control that is necessary over extremely brief time intervals. Video, after all, is just a sequence of many "snapshot" static images. In order to deceive viewers into thinking that they are witnessing continuous motion, each snapshot must be replaced with the next one about 30 times per second. So, at an absolute minimum, creating one second of display on a 1920 × 1080 screen would require 60 million instructions to individual plasma cells, telling each one exactly how long to remain ionized. In fact, the actual number is far higher since, among other things, individual frames are displayed two or more times depending on the set's "refresh rate." So the super-small plasma in each cell must be capable of switching on and off in very short time intervals, and doing so dependably for years.

Withdrawal and Deposits

BUT THE ROLE of plasma physics in microelectronics does not stop there. Controlling those millions of screen instructions, and lurking at the core of every other solid-state device on Earth and above, are various sorts of microchips – arguably the defining technology of the late twentieth century. And increasingly those chips cannot be fabricated without the repeated use of plasmas.

A microchip is built up sequentially, starting from a flat plate of silicon, by adding successive layers of silicon, glass, or metal. At each stage, microscopic features are shaped on each layer, usually employing a "mask" that leaves some areas exposed but covers others as tightly focused laser beams move across the surface. At the end of each shaping operation, the unwanted portion of that layer has to be removed before the next can be added. In the early days of microchip manufacture, this process entailed a liquid chemical wash: after the light beam had altered specific areas on each layer, the layer was bathed in an acid that dissolved away everything except the desired features protected by the mask.

That system worked well when chips were less complex, and features could be larger. For example, Intel's first microprocessor, launched in 1971, had slightly more than 2000 transistors in an area about the size of a child's fingernail. Recently, the same firm rolled out a chip in its Itanium line containing 2 *billion* transistors crammed into approximately the same amount of space.

> A MICROCHIP IS BUILT UP SEQUENTIALLY, AND INCREASINGLY THOSE CHIPS CANNOT BE FABRICATED WITHOUT REPEATED USE OF PLASMAS.

At that dimension scale, liquids are problematic. They can seep under the mask material and are otherwise difficult to control with complete precision as the minimum width of microprocessor components shrinks. (It is now below 50 billionths of a meter and still dropping.) So plasmas came to the rescue.

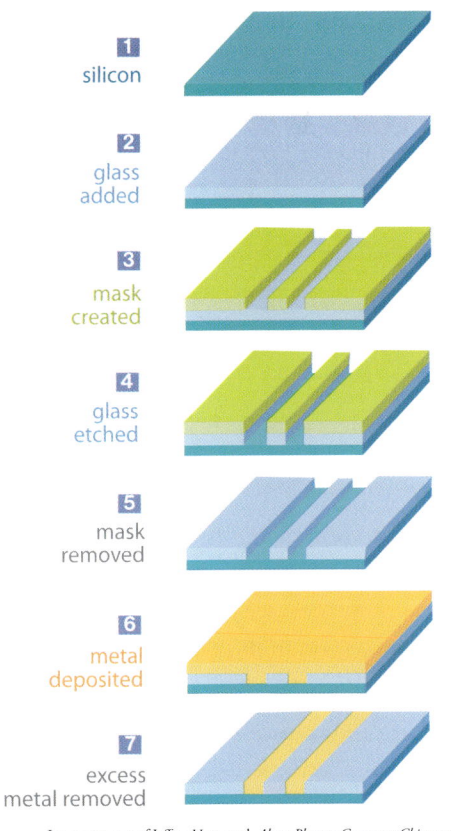

1. silicon
2. glass added
3. mask created
4. glass etched
5. mask removed
6. metal deposited
7. excess metal removed

Image courtesy of Jeffrey Hopwood, *About Plasma–Computer Chips and Plasma* (2006); with permission from the Coalition for Plasma Science.

Because the plasma particles – electrons and ions – are electrically charged, they are far more controllable by electric and magnetic fields than the fluid jumble of atoms in an acid bath. If an appropriate voltage is applied, plasma particles are accelerated straight at their target areas. For example, in order to etch away minuscule portions of a silicon microchip layer, fabricators can position a mask atop the layer and then position it between two oppositely charged metal plates in a vacuum chamber. Then chlorine gas is introduced into the chamber. The high voltage between the plates ionizes the gas, and chlorine ions in the plasma are drawn to the negatively charged silicon sheet. Chlorine is extremely reactive, and the ions eat away unprotected regions of the silicon sheet, producing a silicon–chlorine compound in gas form, which is extracted from the chamber. The result is an exactly carved set of features on a surface that does not have to be rinsed or dried.

Interestingly, plasma techniques are also employed to achieve just the opposite effect – depositing material atop an existing layer. Again, the advantage lies in the ability to fine-tune the interaction between plasmas and adjacent surfaces. Thus, to lay down a new, uniform film of silicon during chip fabrication, the surface is exposed to a gas of a silicon–hydrogen compound called silane (SiH_4). Then the gas is turned to plasma. As the silane molecules break apart, separating some of the hydrogen atoms, silicon–heavy molecules bond to the chip surface. The hydrogen is drawn off with a pump, leaving an even layer of silicon for the next stage in processing.

Many kinds of coatings are applied through plasma technology – including super-hard, diamond-like films that protect surfaces from damage. In this photo, a carbon plasma (the white beam) is sprayed onto a plastic sheet. The sheet is given a very high negative voltage, which accelerates the positively charged carbon ions and imbeds them firmly into the surface. By carefully coordinating the timing of the plasma pulses with the voltage pulses on the plastic, the researchers achieved a high adhesion between ions and surface.

Image courtesy of Blake Wood, Los Alamos National Laboratory.

Plasmas are also employed in many other aspects of microcircuit manufacture, and the lessons learned in their use are now being applied to another leading-edge research field: nanoscience.

Physicists are steadily gaining the ability to manipulate matter at the scales of individual atoms and clusters of atoms, creating assemblies such as carbon nanotubes and nanowires with thicknesses in the range of 10 nanometers (nm, billionths of a meter; a human hair is about 50 000 nanometers wide) and "quantum dots" – microscopic blobs of semiconductor material that contain hundreds or thousands of atoms, but behave collectively like a single individual atom. At those dimensions, matter can behave very differently from its bulk characteristics, displaying novel electrical, optical, magnetic, and mechanical properties that can be exploited to create a new generation of materials, processes, and devices for the twenty-first century.

Plasmas provide a convenient, precisely manageable means of producing these structures. Unlike purely chemical methods, plasma nanoparticle preparation techniques can ensure that particles do not clump together (by giving each one the same charge, so that they repel one another), and can keep them suspended in a reaction for prolonged periods, minimizing the risk of contamination. Plasmas can anneal target substances and synthesize arrangements of atoms that occur only rarely, if ever, under ordinary natural conditions.

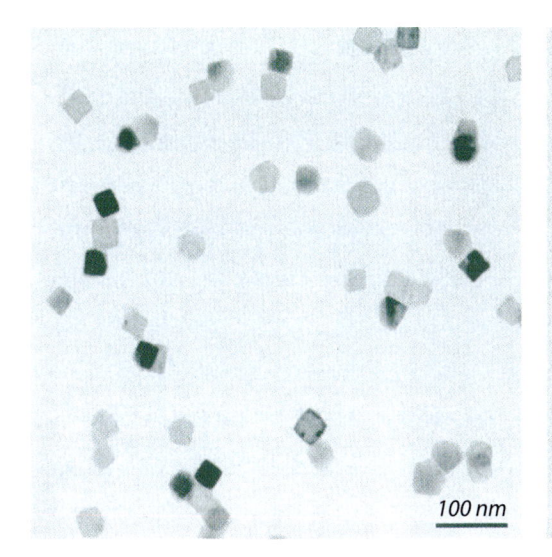

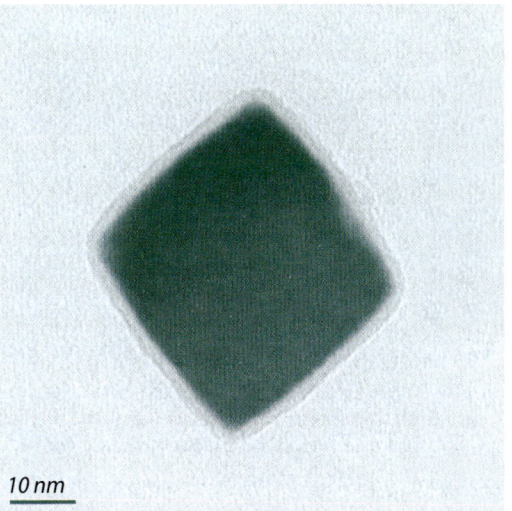

The hugely promising field of nanoscience is revealing the behavior of matter at the dimensions of molecules and atoms. (A nanometer, or one billionth of a meter, is only about a dozen atoms wide.) Low-temperature plasasmas are ideally suited for synthesizing tiny structures such as the silicon nanoparticles shown here, which must be precisely formed in order to have the desired optical and electronic properties.

Image courtesy of A. Bapat *et al.* and the American Institute of Applied Physics.

Plasmas and Human Health

LOW-TEMPERATURE PLASMAS are also of growing importance in many health-related applications. Sterilization and decontamination are among the most prominent. Over the past century, researchers have invented dozens of ways to sterilize objects such as surgical instruments, and the most effective customarily entail powerful chemical fluids and/or high heat.

However, those uses have substantial limitations. They may damage or destroy the very objects they are intended to clean (for example, bio-hazard garments, sensitive detection apparatus, surgically implantable devices), and some of the chemical residue may remain on the objects, causing toxic reactions. Moreover, in the past three decades, scientists have discovered more and more strains of bacteria that can survive – in fact, thrive in — high-temperature environments such as boiling water; and they have found that a few catastrophic conditions, including bovine spongiform encephalopathy, BSE (commonly known as "mad cow disease"), are caused by heat-resistant pathogens that are neither bacteria nor viruses, but misshapen proteins called prions.

These stubborn threats call for new alternatives, and plasma science offers several. One is simple contact with the plasma, which can severely cripple living cells while leaving the sterilized object intact. Another is plasma's ability to produce large numbers of molecular or atomic fragments called "free radicals." These entities have unpaired electrons in their outer shells, making them abnormally reactive. They can wreak chemical havoc on cell membranes or DNA, which is why health-care providers recommend that people consume foods rich in "antioxidants," compounds that interact with radicals before they can do harm. Germs, of course, have no such protection when exposed to free radicals generated by plasmas for decontamination.

These and other plasma effects can be produced at relatively low temperatures, preventing heat damage to the objects being sterilized, and at normal atmospheric pressure, eliminating the need for costly special enclosures. Plasmas can be channeled into streams to clean the inside of narrow tubes such as catheters, and can pervade even the smallest spaces and surface cavities. In addition, as noted in the case of fluorescent lights, certain plasmas are notoriously copious generators of UV radiation, which prevents pathogens from reproducing by breaking up their DNA or RNA.

Finally, low-temperature plasmas can be generated with relatively little power, making them well suited to use on battlefields, remote research sites, and other forbidding conditions.

Plasmas are increasingly used for disinfection and sterilization. These electron microscope images show why. On the left are normal E. coli *microbes. On the right are the damaged remains of those treated with a plasma.*
Image courtesy of M. Laroussi *et al.*, *Applied Physics Letters* **81** (2002), no. 4, pp. 772–774; copyright 2002.

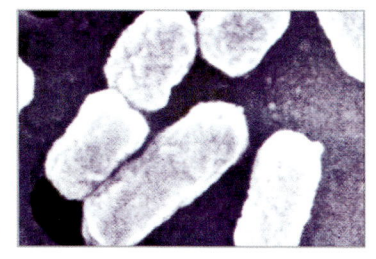

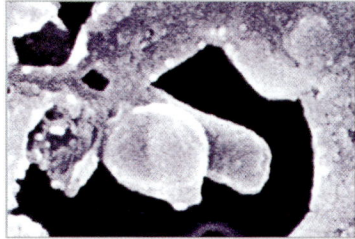

When Push Comes to Shove

THERE ARE NO MORE REMOTE or forbidding conditions than outer space, and that is where low-temperature plasmas display yet another of their many distinctive uses: rocket propulsion, achieved by accelerating plasma ions in electric and magnetic fields.

At first, this notion may seem incomprehensible, since most people associate rocket engines with the ear-splitting chemical roar of combustion as high-pressure gases or solid fuels burn. But silent plasma engines are already routinely used in satellite "station-keeping," the process of boosting a satellite back up to its desired orbit after atmospheric friction causes it to slow and descend. The reason that ion drives work is that all rockets rely on Newton's Third Law of Motion: for every action, there is an equal and opposite reaction. And that principle does not require a flame, or even heat.

Imagine a person, standing motionless, wearing (hypothetically frictionless) roller skates and holding a sack of baseballs. If the skater begins throwing baseballs out in front of him, he will start moving backwards. The more baseballs he throws per unit time, the faster he will go. And, thanks to Newton's Second Law, he can easily substitute any other ball: if he throws all objects with equal speed, then the resulting impulse depends on the total mass ejected. So the skater can achieve about the same speed by throwing three 50-gram golf balls – or thirty 5-gram nickel coins – instead of one 150-gram baseball.

With an electric field of even mediocre strength, ions can be ejected at much higher speed than the exhaust gas of a chemical rocket. Consequently, much less mass of exhaust, and hence of fuel, is needed to accelerate the spacecraft by the same amount. In current plasma-engine designs, ions are brought close to a positively charged anode and propelled away at speeds up to 70 kilometers per second. Xenon is often used as the fuel source, because of its mass (more than twice that of iron) and because it is fairly easy to ionize.

The advantages of plasma engines for space exploration are obvious. They require only 1 percent of the amount of fuel that chemical rocket drives need, and therefore would be ideally suited for long journeys to other planets. In theory, they could eventually use plasmas from fusion reactors, and might scoop up enough free hydrogen in interplanetary space to feed the fusion process continuously.

Plasma-based ion thrusters like this one may propel tomorrow's deep-space probes. Image courtesy of Yevgeny Raitses and Nathaniel Fisch, Princeton Plasma Physics Laboratory, Hall Thruster Experiment.

There are, however, at least three potential drawbacks to ion drives, and plasma researchers have identified promising solutions for two of them. The first is that contact with the plasma can eventually erode away the materials in the exhaust chamber. But by confining the plasma in magnetic fields, most of the particles can be kept away from their surroundings. The second is that an incessant torrent of positive ions can produce a strongly attractive negative charge toward the rear of the device, which pulls the exiting ions backwards. That problem can be solved by engineering a spray of electrons from the back of the drive to neutralize the exhaust jet.

The third problem may be more difficult to solve. In order to generate the electric and magnetic fields necessary for ion drive, an adequate source of electrical power is needed. Photovoltaic systems of the future might suffice in regions near the Sun. But photons are harder to come by as spacecraft move outward in the solar system. So presumably a nuclear power source, or some alternative, will be needed.

With a high degree of confidence, it is likely that these challenges – and scores of others in physics, engineering, and technology – will be met and overcome by ongoing progress in plasma science. Only 80 years have passed since the term "plasma" was first used in physics, and modern practice is barely a half century old. Indeed, the Division of Plasma Physics within the American Physical Society did not exist until 1958.

Yet in those 50 years, researchers have plumbed the mysteries of the Sun, trapped 100-million-degree plasmas in magnetic bottles, developed accurate mathematical models of a host of wave and particle phenomena, mapped the shape and structure of Earth's churning magnetosphere, and extended plasma science to the titanic forces and fields at the outermost reaches of the cosmos. There is every reason to believe that understanding will continue to advance, as physics reveals more of the remarkable roles that plasmas play in nature.

> ONLY 80 YEARS HAVE PASSED SINCE THE TERM "PLASMA" WAS FIRST USED IN PHYSICS…IN THOSE YEARS RESEARCHERS HAVE PLUMBED THE MYSTERIES OF THE SUN, TRAPPED 100-MILLION-DEGREE PLASMAS IN MAGNETIC BOTTLES, DEVELOPED ACCURATE MATHEMATICAL MODELS, MAPPED THE SHAPE AND STRUCTURE OF EARTH'S CHURNING MAGNETOSPHERE, AND EXTENDED PLASMA SCIENCE TO THE OUTERMOST REACHES OF THE COSMOS.

Index

Figures and illustrations are indicated by *italic page numbers*; footnotes by the suffix "*note*"

accretion discs 56
   around black holes 56
active galactic nuclei 56, 62
active regions (on Sun) 30
   flares and 32
air, as non-conductor 5
Alfvén, Hannes 7, 20, 25
Alfvén waves 20–4, 25, 30
alpha-particle energy 23
American Physical Society, Division of Plasma Physics 72, xi
antioxidants 70
Appleton, Edward 7
The Arches region *51*
Arrhenius, Svante 5
astrophysics 7
atomic structure 2, 5
auroras 1, *3*, *4*, 36
   at Jupiter's poles 37

bacteria, heat-resistant 70
"balanced charges" concept 9
Big Bang theory 51
binding energy (for nucleons) 27
black holes 56
   accretion discs around 56–8
bovine spongiform encephalopathy (BSE) 70

carbon coatings 68
"cathode rays" 5
characteristics of plasmas 9, 11, 12, 43
chromosphere (of Sun)
   Alfvén waves in 23, 30
   loops in 30
coating technology 68
Coma Cluster of galaxies 62
comet tails *26*, 34
convection processes 28
corona (of Sun) 21, 30
   see also solar corona
coronal mass ejections (CMEs) 33
   effects 33, 37
cosmic rays 60–2
   energy range 60, *61*

   number arriving per unit area *61*
   sources 60, 62
Crab Nebula *54*
current–magnetic field interactions 12, 28

Debye length 11
decontamination 70
definition (of plasmas) 1, 9, 11
desktop particle accelerators 8, 17–18
deuterium (hydrogen isotope) 39, 40
DIII-D tokamak *6*, *44*
dynamos 52

Earth
   ionosphere 2, *3*, *4*
   magnetic field 53
   magnetic polarity reversal 52
   magnetosphere 7, 36
   molten core 52
Einstein's mass–energy equation 18
electrical conductivity of plasmas 5, 12
electromagnetic (EM) waves 20
   propagation in plasma 20, 25
electron volt 2
electrons 5
   compared with protons 12
electrostatic probe 10
event horizon (of black hole) 56
everyday applications of plasmas 1, 2, 4, 63–9
examples (of plasmas) *4*

Faraday, Michael 5
Fermi, Enrico 25, 62
Fermi acceleration 62
flames, plasmas in *4*
flat-screen TVs 66
fluorescent lights 2, 63, 64–5
   plasma temperature in *4*, 64
"fourth state of matter" 1
free radicals 70
frictional heating in accretion discs 56

fusion reactor(s)
    Alfvén waves in 23
    nuclear reactions in *39*, 40, 43–4
    temperatures in 2, *4*
    see also inertial confinement fusion reactor(s); magnetic confinement fusion reactor(s)
fusion research 49

galactic black holes 56
galaxies, magnetic fields 52–3
gamma-ray bursters 56, 62
gamma rays, emission from neutron stars 55, 56
geodynamos 52–3
geomagnetic storms 37
geoscience 7
gravitational potential energy 56

Hale, George Ellery 29, 30
harmonic oscillator 16
health-related applications 70
Heaviside, Oliver 6
Helios solar orbiter 17
helium nuclei 23, 39, *40*, 43
helium plasma, Alfvén waves in 21, *22*
high-energy-density physics (HEDP) 49
high-intensity discharge lighting units 65
Hinode ("Sunrise") solar orbiter observations 23, 29, 30–1
hohlraum 8, 46, *47*
Hubble Space Telescope observations 37, 57
hydrogen atom, electron in 2

inertial confinement (IC) fusion reactor(s) 46–8
    compared with magnetic confinement devices 47
    direct-drive method 46
    experiments 47
    "fast ignition" approach 48
    indirect-drive method 46, 47
    temperature range *4*
    Z-inch configuration *14*, 49
Intel microprocessors 67
International Thermonuclear Experimental Reactor (ITER) *42*
interstellar space, plasmas in 1, 2, *4*
ion drives/thrusters 71–2
    drawbacks 72
ionized gas 1
    Langmuir's studies 6, 10
ionosphere 2, *3*, 4
ions 5

jets from accretion discs 59
Jupiter's magnetic field 37

Kelvin temperature scale 2*note*
Kennelly, Arthur 6
Kepler's laws 57

lamps 2, *15*, 63, 64–5
    see also fluorescent lights
Langmuir, Irving 6, 10
Langmuir waves 16–17
Large Hadron Collider (LHC) 18, 60
Large Plasma Device (LAPD) 21
laser fusion 8, 46
lasers 7–8
    in inertial confinement fusion reactor(s) 46–8
lightning 1, *4*, 5, 12, *13*
low-temperature plasmas 70

M87 galaxy, jet from 59
magnetars 55
magnetic ballast (in fluorescent light) 64–5
magnetic confinement fusion reactor(s) 6, 41–5
    Alfvén waves in 23
    compared with inertial confinement devices 47
    plasma defects 44–5
    temperature range *4*
    see also tokamak(s)
magnetic field–current interactions 12, 28
magnetic islands in fusion plasmas 44–5
magnetic mirror 62
magnetic reconnection *32*, *33*, 45
magnetohydrodynamics 7, 25
magnetorotational instability 58
magnetosphere 7, 36
    penetration by solar-wind particles 36, 37
magnetotail 37
Marconi, Guglielmo 6
mass–energy equivalence equation 18, 28
microchip manufacture 67–8
    liquid chemical process 67
    plasmas used 2, 68
momentum conservation in black hole accretion disc 57–8

nanoflares (in Sun) 30
nanometer, meaning of term 69
nanoparticle preparation techniques 69
nanoscience 69
National Ignition Facility 48
    laser bays *48*

nebulas 54
neon signs *4*
neutral beam heating (in tokamak) 43
neutron stars 54–5
neutrons
    binding to protons 27
    in deuterium and tritium nuclei 39
    formation in collapsing supernova 54
    release in nuclear fusion 39, *40*, 43, 44
Newton's Second Law 71
Newton's Third Law 46, 71
NGC 4261 galaxy *57*
non-neutral plasmas 9
Northern Lights *3*, 33
    see also auroras
nuclear fission 39
nuclear fusion 39
    advantages 40
    power generation using 7, 39–50
    in Sun (and other stars) 27, 40
    in weapons 7, 40
    see also fusion reactors
nucleons 27
    combining 27–8

ohmic heating (in tokamak) 43

Parker, Eugene N. 34, 38
particle colliders 18
photosphere (of Sun) 21, 29
pixels 66
plasma engines 71
plasma frequency 16
plasma lamp *15*
plasma oscillations 16–19
plasma sheaths *11*
plasma sheet (in solar wind–magnetosphere interaction) 37
plasma TV screen 2, 66
plasma wake *19*, *26*
plasma welding 10
polarity in plasmas 9, 11
poloidal magnetic fields
    black holes and 58
    in fusion reactors 41
prions 70
protons 27
    compared with electrons 12
pulsars 55

quantum dots 69
quasi-neutral plasmas 9

radio communications 6–7
radio-frequency heating (in tokamak) 43
rocket propulsion 71
Rosenbluth, Marshall 50

shielding effects 9, 11
shock acceleration, cosmic rays affected by 62
silane 68
silicon depositing 68
silicon etching 67, 68
SOHO (Solar and Heliospheric Observatory) measurements 28
solar core, temperature in *4*, 27
solar corona 21, 30
    Alfvén waves and 23, 30
    Langmuir waves 16–17
    temperature *4*, 21, 30, 34
solar cycle (maximum/minimum activity), solar wind and *34*, 37
solar energetic particles 60
solar flares 17, 30, 32–3
    effects 33
solar loops 30
solar prominences *23*
solar wind 2, *3*, *4*, 24, 34–7
    confirmation of existence 35, 38
    first proposed 34, 38
    interaction with Earth's upper atmosphere *3*, 36
    interaction with Venus atmosphere 35
    turbulence in 24
South Atlantic Anomaly (Van Allen radiation belt) 36
space exploration 71
stars, number in Galaxy 27
sterilization 70
"strong" interaction 27
sub pixel 66
Sun
    convection processes 28
    magnetic field 29, 53
    magnetic polarity reversal 29, 52
    rotational rate 28
    see also solar...
sunspots 29–30
    flares and 32
    observations *30–1*
supernovas 54

tabletop particle accelerators 8, 17–18
temperature scales 2*note*
Thomson, J.J. 5
tokamak(s) *41*, 42
    "current drive" method 42
    firing up (temperature raising) techniques 43
    see also magnetic confinement fusion reactor(s)
toroidal magnetic fields
    black holes and 58
    in fusion reactors 41
TRACE (Transition Region and Coronal Explorer)
    observations *31*, *32*
transport barriers in fusion reactor
    plasma 45
tritium (hydrogen isotope) 39, 40
turbulence
    plasma in fusion reactor 45
    solar wind 24
TV screens 2, 66
    resolution 66

ultraviolet (UV) radiation
    effects 70
    in fluorescent lights 64, *65*
    in TVs 66

vacuum tube *11*
Van Allen, James 7
Van Allen radiation belts 7, 25, 36
Venus's atmosphere, interaction with solar wind *35*
video monitors 66

wakefield accelerators 17–18
weapons 7
word "plasma" first used in physics 6, 72

X-rays
    emission from black hole accretion discs 56
    emission from neutron stars 55

Z-pinch inertial confinement devices *14*, 49